Advanced Computational Vibroacoustics

Advanced Computational Vibroacoustics presents an advanced computational method for the prediction of sound and structural vibrations in low- and medium-frequency ranges of complex structural acoustics and fluid-structure interaction systems encountered in aerospace, automotive, railway, naval, and energy-production industries. The formulations are presented within a unified computational strategy and are adapted for the present and future generation of massively parallel computers.

A reduced-order computational model is constructed using the finite element method for the damped structure and the dissipative internal acoustic fluid (gas or liquid with or without free surface) and using an appropriate symmetric boundary-element method for the external acoustic fluid (gas or liquid). This book allows direct access to computational methods that have been adapted for the future evolution of general commercial software. Written for the global market, it is an invaluable resource for academic researchers, graduate students, and practicing engineers.

Roger Ohayon joined Conservatoire National des Arts et Métiers as Professor Chair of Mechanics, where he is now emeritus professor, after completing research at Office National d'Etudes et Recherches Aérospatiales, the aerospace research laboratory in France. He is a Fellow of several associations, including AIAA, ASME, and IACM, and he is the recipient of the Gay-Lussac Humboldt Research Award, the SPIE's lifetime achievement award, the ASMS/ASME/AIAA Award, the Prandtl Award from Eccomas, several IACM Awards, the EASD Senior Prize, and the French Academy of Sciences Award. His expertise is in mechanical and computational modeling of fluid-structure interaction problems, structural acoustics, and smart structural systems. He is on the editorial board of thirteen international journals, including the *International Journal for Numerical Methods in Engineering*, *Computer Methods in Applied Mechanics and Engineering*, and *Computational Mechanics*, and he is the associate editor of the *Journal of Intelligent Material Systems and Structures* and the *AIAA Journal*. He is the coeditor of several books and the coauthor of more than 100 publications in refereed international journals. He is the coauthor of two books, *Fluid-Structure Interaction* (with H. J.-P. Morand, 1995) and *Structural Acoustics and Vibration* (with Christian Soize, 1998).

Christian Soize joined Université Paris-Est Marne-la-Vallée (UPEM/Modeling and Multiscale Simulation Research Laboratory) as Professor in Mechanics, after completing research at Office National d'Etudes et Recherches Aérospatiales. He is a Fellow of the Acoustical Society of America and has received a number of awards and honors, including the Senior Research Prize from EASD, a research award from the International Association for Structural Safety and Reliability, and the Noury Prize from the French Academy of Sciences. He is the author or coauthor of more than 170 papers in refereed international journals and of 7 books, including *Mathematics of Random Phenomena* (with P. Krée, 1986), *The Fokker-Planck Equation for Stochastic Dynamical Systems and Its Explicit Steady State Solutions* (1994), *Structural Acoustics and Vibration* (with R. Ohayon, 1998), and *Stochastic Models of Uncertainties in Computational Mechanics* (2012). He has pioneered a number of new approaches in stochastic modeling of complex systems, including fuzzy structure theory, the concept of an energy operator for dynamics in the medium-frequency range, and, more recently, the concept of a nonparametric probabilistic approach for model uncertainties in computational mechanics and vibroacoustics.

ADVANCED COMPUTATIONAL VIBROACOUSTICS

Reduced-Order Models and Uncertainty Quantification

Roger Ohayon

Conservatoire National des Arts et Métiers

Christian Soize

Université Paris-Est

CAMBRIDGE
UNIVERSITY PRESS

CAMBRIDGE
UNIVERSITY PRESS

32 Avenue of the Americas, New York, NY 10013-2473, USA

Cambridge University Press is part of the University of Cambridge.

It furthers the University's mission by disseminating knowledge in the pursuit of education, learning, and research at the highest international levels of excellence.

www.cambridge.org
Information on this title: www.cambridge.org/9781107071711

First published 2014

Printed in the United States of America

A catalog record for this publication is available from the British Library.

Cambridge University Press has no responsibility for the persistence or accuracy of URLs for external or third-party Internet Web sites referred to in this publication and does not guarantee that any content on such Web sites is, or will remain, accurate or appropriate.

CONTENTS

PRINCIPAL OBJECTIVES AND
A STRATEGY FOR MODELING
VIBROACOUSTIC SYSTEMS

In this book, we are interested in the analysis of *vibroacoustic systems*, which are also called *structural acoustic systems* or *fluid-structure interactions for compressible fluid* (gas or liquid). *Vibroacoustics* concerns noise and vibration of structural systems coupled with external and/or internal acoustic fluids. *Computational vibroacoustics* is understood as the numerical methods solving the equations of physics corresponding to vibroacoustics of complex structures. *Complex structures* are encountered in many industries for which vibroacoustic numerical simulations play an important role in design and certification, such as the aerospace industry (aircrafts, helicopters, launchers, satellites), automotive industry (automobiles, trucks), railway industry (high speed trains), and naval industry (ships, submarines), as well as in energy production industries (electric power plants).

Since we are interested in the analysis of general complex structural systems in the sense of computational methods defined here, we do not consider analytical or semianalytical methods devoted to structures with simple geometry, asymptotic methods mainly adapted to the high-frequency range (statistical energy analysis, diffusion of energy, etc.) and approaches that imply them. Concerning the latter, the coupling of the local dynamic equilibrium equation (finite element method) and power balances (implemented in the spirit of the statistical energy analysis) have been analyzed in Soize (1998); Shorter and Langley (2005); Cotoni et al. (2007). Nevertheless, this kind of approach is not purely "computational" in the sense described earlier and requires specific hypotheses concerning certain subsystems that are supposed to have a high-frequency dynamic behavior (high modal density).

By *advanced computational acoustics*, we mean computational methods that are adapted to the present and future generation of massively parallel computers and for which the formulations are adapted to analyze complex structural acoustic systems, and which will require a huge number of degrees of freedom (several billion) in order to correctly model structures with complex geometries and made up of different materials having complex microstructures such as composites, metamaterials and acoustic coating such as foams for soundproofing.

This book does not pretend to be a review of the existing approaches and methodologies devoted to the vibroacoustic field. Physical bases useful for vibroacoustics can be found in numerous books, such as in Morse and Ingard (1968); Cremer et al. (1988); Pierce (1989); Crighton et al. (1992); Landau and Lifchitz (1992); Junger and Feit (1993); Morand and Ohayon (1995); Ohayon and Soize (1998); Blackstock (2000); Bruneau (2006); Fahy and Gardonio (2007).

The book is relatively short in order to allow readers interested in computational vibroacoustics of complex structural systems to directly access computational methods that are adapted to the present and future evolutions of the general commercial softwares.

Finally, it should be noted that, obviously, a short book cannot include all the various methodologies in computational vibroacoustics of complex structural systems (but which have been mentioned and referenced throughout the text). More specifically, this short book proposes a unified strategy chosen by the authors and gives their view in this field in the context of the present and future generations of massively parallel computers. The methods proposed have effectively been validated and applied to complex structural systems and can partly be found in the references.

In the first section, the principal objectives of the book are presented. The next sections are devoted to a strategy for modeling complex vibroacoustic systems adapted to computational vibroacoustics. This strategy will be the guideline to be followed in subsequent chapters to present advanced computational vibroacoustics.

1.1 PRINCIPAL OBJECTIVES OF THE BOOK

The fundamental objective of this book is to present an advanced *computational method* for the prediction of the responses in the frequency domain of general *linear vibroacoustic systems*. The frequency domain is usually composed of three parts: the *low-*, the *medium-* and the *high*-frequency ranges which are defined in the next section. This book is devoted to computational vibroacoustics in the low- and medium-frequency ranges.

The system under consideration is constituted of a deformable *dissipative structure*, coupled with an *acoustic cavity*. This cavity is filled with a *dissipative acoustic fluid*, which is an inviscid acoustic fluid for which a damping model is introduced. In addition, *wall acoustic impedances* can be taken into account, which allow us to model the acoustic properties of the physical wall constituting a part of the geometrical interface between the internal acoustic fluid and the structure. The system is surrounded by an inviscid acoustic fluid occupying an infinite domain, called *external acoustic fluid*, which is coupled with the structure through the geometrical interface between the external acoustic fluid and the structure.

The vibroacoustic system is submitted to given *acoustic sources* (such as loudspeakers) in the cavity (internal acoustic sources) and in the external acoustic fluid domain (external acoustic sources), as well as given *mechanical forces* applied to the structure (such as surface and body forces).

The frequency responses of the vibroacoustic system are the *displacement field* in the structure, the *pressure field* in the acoustic cavity, and the *pressure fields* on the external fluid-structure interface and in the external acoustic fluid (near and far fields).

It is now well established that the predictions in the medium-frequency range must be improved by taking into account both the *system-parameter uncertainties* and the *model uncertainties* induced by *modeling errors*. Such aspects will be considered in Chapter 9, devoted to *uncertainty quantification* (UQ) in vibroacoustics (structural acoustics) and in fluid-structure interaction.

In this book, the presented formulations, which correspond to new extensions and complements with respect to the state-of-the-art, can be used for the development of a new generation of computational vibroacoustic softwares in particular, in the context of parallel computers. We present here an advanced computational formulation that is based on an efficient reduced-order model in the frequency range and for which all the required modeling aspects for the analysis of the low- and medium-frequency ranges have been taken into account. More precisely, we have introduced a frequency-dependent linear constitutive equation for modeling damping effects in complex structures, an appropriate dissipative model for the internal acoustic fluid including wall acoustic impedance, and, finally, a global model of uncertainty. It should be noted that model uncertainties must absolutely be taken into account in the computational models of complex vibroacoustic systems in order to improve the prediction of the responses in the low- and medium-frequency ranges.

The reduced-order computational model is constructed using the finite element method for the structure and for the internal acoustic fluid. The external acoustic fluid is treated using an appropriate boundary element method in the frequency domain.

Throughout the book, the finite element method (FEM) (see, for instance, Hughes, 2000; Zienkiewicz and Taylor, 2005) is used for the spatial discretization of the boundary value problems yielding the associated matrix equations. An alternative method to construct the matrix equations would consist in using the isogeometric analysis (see Hughes et al., 1996).

For some details of mathematical developments presented in this book, the reader is referred to Ohayon and Soize (1998), and for a brief general overview of the recent aspects, the reader is referred to Ohayon and Soize (2012).

In this book, we do not consider fluid flows (for fluid dynamics, see Batchelor 2000), such as the case of a structure in movements in an external fluid at rest (or a the case of a structure at rest in a flow) and the case of a structure with internal flows. For physical and modeling aspects related to flows in acoustics and fluid-structure interactions, we refer the reader to Howe (2008). For fluid-structure interaction with internal flow using nonlinear computational fluid dynamics, we refer the reader to Bazilevs et al. (2013).

Therefore, the considered modeling is carried out for an equilibrium state for which the acoustic fluid is at rest. Nevertheless, such a modeling can be used, for instance, for external flows around the structure or for internal flows inside pipes when the aeroelastic or the hydroelastic phenomena are decoupled from the vibroacoustic phenomena under consideration. Therefore, only the effects of the external flow in terms of the forces applied to the structure are considered; this is the case of the effects induced by a turbulent boundary layer on the structure.

1.2 DEFINITION OF THE DIFFERENT FREQUENCY RANGES: LF, MF, AND HF

The different types of vibration responses of a weakly dissipative complex structure lead us to define the frequency ranges of analysis. Let $u_j(\mathbf{x}, \omega)$ be the frequency response function (FRF) of a component j of the displacement $\mathbf{u}(\mathbf{x}, \omega)$, at a fixed point $\mathbf{x}$ of the structure and at a fixed circular frequency ω (in rad/s). Figure 1.1 represents the modulus $|u_j(\mathbf{x}, \omega)|$ of the FRF in log scale and Figure 1.2 represents the unwrapped phase $\varphi_j(\mathbf{x}, \omega)$ of the FRF such that $u_j(\mathbf{x}, \omega) = |u_j(\mathbf{x}, \omega)| \exp\{-i\varphi_j(\mathbf{x}, \omega)\}$. The unwrapped phase is defined as a continuous function of ω obtained in adding multiples of $\pm 2\pi$ for jumps of the phase angle. Three frequency ranges can then be characterized, as follows.

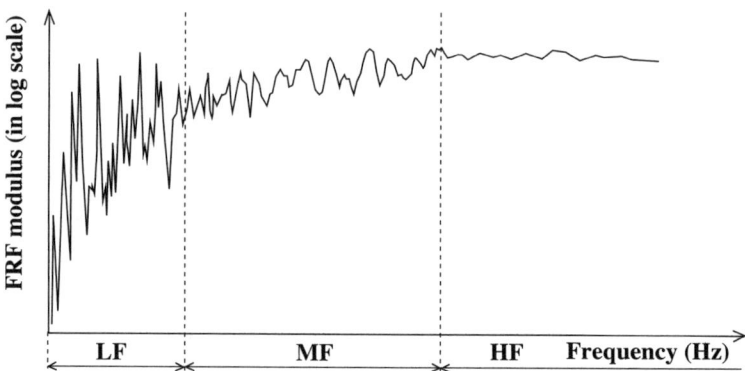

Figure 1.1 Modulus of the FRF as a function of the frequency. Definition of the LF, MF, and HF ranges.

(1) The *low-frequency range* (LF) is defined as the modal domain for which the associated conservative system exhibits isolated modes (low modal density). In this LF range the modulus of the FRF exhibits isolated resonances whose amplitudes are driven by the damping (see Figure 1.1) and the phase rotates of π at the crossing of each isolated resonance (see Figure 1.2).

(2) The *high-frequency range* (HF) is defined as the frequency band for which there is a high modal density that is approximatively constant

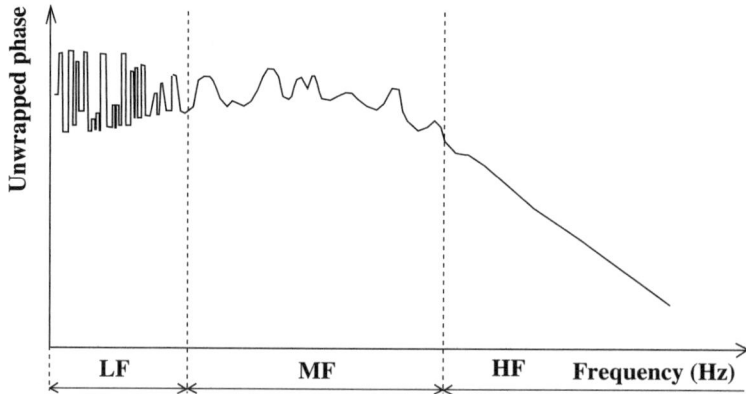

Figure 1.2 Unwrapped phase of the FRF as a function of the frequency. Definition of the LF, MF, and HF ranges.

as function of the frequency. Such an assumption is required to use asymptotic approaches and/or statistical descriptions (as carried out by the statistical energy analysis). In this HF range the modulus of the FRF varies slowly as the function of the frequency (see Figure 1.1) and the phase is approximatively linear (see Figure 1.2). It should be noted that this frequency range is not defined with respect to the absolute value of the frequency but is mainly related to the modal density of the system. This frequency range is outside the scope of this book.

(3) For a complex structure (complex geometry, heterogeneous materials, complex junctions, complex boundary conditions, several attached equipments, etc.), an intermediate frequency range called the *medium-frequency range* (MF) appears. This MF range does not exist for a simple structure (for example, a simply supported homogeneous straight beam). This MF range is defined as the intermediate frequency band for which the modal density exhibits large variations over the band. Contrary to the LF range, due to the effects of the damping, the frequency response functions do not exhibit isolated resonances (see Figure 1.1) and the phase slowly varies as a function of the frequency (see Figure 1.2).

This analysis presented for a structure can be extended for a complex vibroacoustic system. This book is devoted to LF and MF ranges analyzes for complex vibroacoustic systems.

1.3 STRATEGY FOR MODELING COMPLEX VIBROACOUSTIC SYSTEMS

Computational vibroacoustics is devoted to the computation of the responses of a vibroacoustic system (which can be a macrosystem or a microsystem). This vibroacoustic system is constituted of a deformable structure made up of metallic materials, heterogeneous composite materials, and, more generally, metamaterials. Let us also mention investigations of new types of materials, called metamaterials, for cloaking purposes connected to acoustic anechoicity (Milton et al., 2006; Fang et al., 2006; Chen and Chan, 2007; Pendry and Li, 2008; Cheng et al., 2008). It should be noted that the class of materials considered for the structure can be extended for the meta-smart-adaptive heterogeneous materials but are outside the scope of the book. The structure is coupled with an internal acoustic cavity and an external acoustic fluid. Acoustic coatings (soundproofing materials, sound-insulation layers, etc.), made up, for instance, with cellular materials such as porous materials, can be taken into account on the fluid-structure interfaces.

The responses of the vibroacoustic system are mainly driven by the structure and possibly by the internal acoustic cavity which are resonant systems, but not by the infinite external acoustic fluid which is not a resonant system.

(i) In the first step, let us precise the objectives of the vibroacoustic modeling, which consist of computing the responses of the vibroacoustic system submitted to prescribed excitations.
   - The excitations are forces applied to the structure (forces generated by the environment and transmitted to the structure by solid paths such as shocks or vibration equipments or through fluid paths such as turbulent boundary layer effects), internal acoustic sources inside the acoustic cavity and external acoustic sources.
   - The responses are the structural displacement field, the pressure field in the internal acoustic cavity for internal acoustic noise quantification, and the near and the far pressure fields in the external acoustic fluid. Those pressure fields can be decomposed as the sum of
     - an incident field due to external acoustic sources,
     - a scattering of the incident acoustic field by the rigid structure,

    ○ a radiation acoustic field due to the deformation of the structure.

The so-called backscattered acoustic field is the sum of the scattering and radiation fields.

- The formulation presented in the book also includes the following cases:
  - ○ the anechoicity analysis, which consists in studying the backscattered acoustic field of an incident acoustic field by the structure;
  - ○ the acoustic stealth of the structure, which consists in analyzing the transmission into the external acoustic fluid, through the structure, of internal acoustic sources located inside the acoustic cavity or of mechanical forces applied to the structure.

(ii) In the second step, let us specify that the modeling strategy is based on a formulation in the frequency domain. In effect, there are many advantages to using a frequency domain formulation instead of a time domain formulation for linear vibroacoustic systems. The first one is the possibility to analyze the vibroacoustic system in terms of the frequency ranges, that is, to consider the low-, the medium-, and the high-frequency ranges. The second one is the capability to choose the vibroacoustic system modeling accordingly to frequency range. In particular, the damping models and the quantification of uncertainties must be taken into account in the medium-frequency range. Finally, the formulation in the frequency domain can easily be implemented in massively parallel computers because the calculations can be distributed frequency by frequency.

It should be noted that the low-frequency range is mainly driven by the resonances induced by the elastic modes of the structure, by the acoustic modes of the acoustic cavity, and by the elastoacoustic modes of the vibroacoustic system. Concerning the appropriate formulations for computing the elastoacoustic modes of the associated conservative vibroacoustic system, including substructuring techniques, and for the construction of the corresponding reduced-order model, we refer the reader to Morand and Ohayon (1995); Ohayon et al. (1997); Ohayon and Soize (1998); and Ohayon (2004a,b) (including the references).

(iii) In the third step, the strategy concerning the damping modeling is presented for the vibroacoustic system, as follows.

- *External acoustic fluid.* The viscosity of the fluid is negligible in the considered low- and medium-frequency ranges for usual acoustic fluids such as air or water. Therefore the acoustic fluid is inviscid and there is no damping inside the fluid. Nevertheless, since the external acoustic fluid occupies an infinite domain, the outward Sommerfeld radiation condition at infinity implies that the energy traveling towards infinity does not come back and consequently, corresponds to an energy loss. Due to the coupling between the structure and the external acoustic fluid, this phenomenon yields an apparent damping for the structure because the mechanical energy transmitted by the structure to the external acoustic fluid is lost for the structure.

- *Internal acoustic fluid.* The acoustic fluid in the cavity is dissipative and a damping model must be introduced. Generally, there are two main physical dissipations. The first one is an internal acoustic dissipation inside the cavity due to the viscosity and the thermal conduction of the fluid. The second one is the dissipation generated inside the wall viscothermal boundary layer of the cavity. In general the first one is negligible with respect to the second one. However, we propose to simply take into account the two sources of damping by introducing an equivalent damping term corresponding to an internal acoustic dissipation inside the volume of the cavity.

- *Structure.* For viscoelastic materials, a frequency dependent linear constitutive equation is introduced using the linear theory of viscoelasticity. For elastic materials, a frequency dependent linear constitutive equation is introduced with a linear viscous damping term and/or with a parameterized family of frequency dependent damping models.

- *Internal fluid-structure interface.* The acoustic damping properties of the physical wall, which constitutes a part of the geometrical interface between the internal acoustic fluid and the structure, are due to the acoustic properties of the material constituting the interface (material of the structure or acoustic coating material embedded in the structure or in the vicinity of its boundary). This type of acoustical properties on the geometrical interface will be modeled by introducing a local wall acoustic impedance. In addition, if an acoustic coating is placed on a part of the internal fluid-structure interface, two types of modeling can be used.

  - The first one consists of considering the coating material as a material embedded in the structure. In such a case, this acoustic coating is a part of the structure and is then modeled through the structural modeling.

  - The second one consists of constructing an equivalent model of the coating material by introducing a wall acoustic impedance on the interface.

- *External fluid-structure interface.* If coating materials are placed on the external fluid-structure interface, these materials are assumed to be embedded in the structure and, therefore, are then modeled through the structural modeling. It should be noted that for such coatings, a wall acoustic impedance model is not used.

(iv) Finally, we summarize hereinafter the *chosen strategy for the advanced computational vibroacoustics* that is described throughout the book:

- Symmetric boundary element method (BEM) without spurious frequencies for the external acoustic fluid (Chapters 3 and 10).

- Finite element method (FEM) for internal dissipative acoustic fluid with wall acoustic impedance (Chapters 4 and 7).
- Structural modeling using frequency dependent linear constitutive equations for modeling damping effects in complex structures (Chapters 5 and 7).
- Reduced-order computational vibroacoustic model (ROM) (Chapter 8, based on Chapters 6 and 7).
- Uncertainty quantification (UQ) (Chapter 9).

2

DEFINITION OF THE
VIBROACOUSTIC SYSTEM

As explained in Chapter 1, we consider a mechanical system constituted of a linear damped structure Ω_S containing a dissipative acoustic
fluid (gas or liquid) which occupies a domain Ω. This system is surrounded by an infinite external inviscid acoustic fluid domain Ω_E (gas
or liquid) (see Figure 2.1). A part Γ_Z of the internal fluid-structure
interface is assumed to be dissipative and is modeled by a wall acoustic local impedance Z. This system is submitted to a given internal
acoustic source in the acoustic cavity and to given mechanical forces
applied to the structure. In the external acoustic fluid domain, external acoustic sources are given. The structure is assumed to be in a
free-free condition, which means that the structure, considered as a
three-dimensional medium, possesses six rigid body motions (three
translations and three rotations), which is the general case and which
corresponds to many practical situations (aircrafts in flight, ships, space
launch vehicles in flight, etc.). However, for sake of brevity, we then
do not consider here the particular cases for which the structure is
fixed or submitted to prescribed displacements. It should be noted
that those particular situations can easily be taken into account in the
computational vibroacoustic model that is described in this book.

The given mechanical forces and the other forces acting on the
structure which are induced by external and internal acoustic sources,
that is, the external forces applied to the structure, are assumed to
be in equilibrium (the resultant force and the corresponding moment

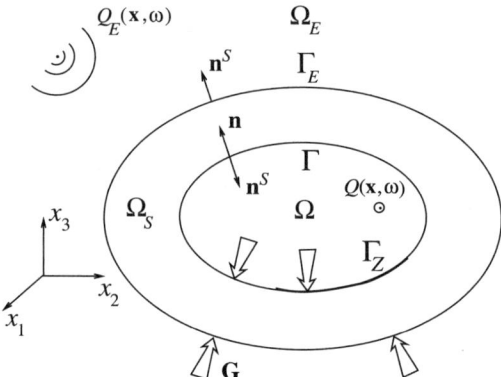

Figure 2.1 Configuration of the vibroacoustic system.

with respect to any point are zero). In this chapter, the vibroacoustic system is described and the mechanical and acoustical hypotheses are introduced. The vibroacoustic system is assumed to be in linear vibration around a static equilibrium configuration taken as a natural state at rest. This means that the prestresses are not taken into account, as done for the finite element model of the structure in vacuo for which prestresses are classically introduced through a geometric stiffness matrix (see Zienkiewicz and Taylor, 2005). This chapter is devoted to the description of the vibroacoustic system and to the mechanical and acoustical hypotheses. The first section deals with the notations. In a second section, the hypotheses are introduced for the structure. The internal dissipative acoustic fluid is then introduced. Finally, the last section is devoted to the external inviscid acoustic fluid.

2.1 NOTATIONS

The physical space is referred to a cartesian reference system and the generic point is denoted as $\mathbf{x} = (x_1, x_2, x_3)$. Throughout the book, the classical convention is used for summations over repeated Latin indices, but not over Greek indices. If $\mathbf{a} = (a_1, a_2, a_3)$ and $\mathbf{b} = (b_1, b_2, b_3)$ are two vectors, the inner product (or scalar product) of

$\mathbf{a}$ and $\mathbf{b}$ is defined by $\mathbf{a} \cdot \mathbf{b} = a_j b_j$. For any function $f(\mathbf{x})$, $f_{,j}(\mathbf{x})$ denotes the partial derivative $\partial f(\mathbf{x})/\partial x_j$ with respect to x_j. The gradient operator ∇f is the vector $(f_{,1}, f_{,2}, f_{,3})$ and the three-dimensional-Laplacian operator is such that $\nabla^2 f = f_{,jj}$.

As explained earlier, linear vibration problems are formulated in the frequency range for vibroacoustic systems. The Fourier transform (Papoulis, 1977; Pinkus and Zafrany, 1997) is a mathematical tool that allows time domain problems to be transformed in frequency domain problems. Therefore, the following notation is introduced for the Fourier transform of various time functions involved in the book. Concerning the structure, for the time-dependent displacement field $\mathbf{u}(\mathbf{x}, t) = (u_1(\mathbf{x}, t), u_2(\mathbf{x}, t), u_3(\mathbf{x}, t))$, for the time-dependent strain $\varepsilon_{ij}(t)$ and stress $\sigma_{ij}(t)$ tensors at a given point $\mathbf{x}$ (in the following, $\mathbf{x}$ is removed from $\varepsilon_{ij}(t)$ and $\sigma_{ij}(t)$), we will introduce the following simplified notation consisting in using the same symbol for a quantity and its Fourier transform. Let i be the pure imaginary complex number. For all real values ω (in rad/s) of the circular frequency, we have

$$\mathbf{u}(\mathbf{x}, \omega) = \int_{-\infty}^{+\infty} e^{-i\omega t} \, \mathbf{u}(\mathbf{x}, t) \, dt \,, \tag{2.1}$$

$$\varepsilon_{ij}(\omega) = \int_{-\infty}^{+\infty} e^{-i\omega t} \, \varepsilon_{ij}(t) \, dt \,, \tag{2.2}$$

$$\sigma_{ij}(\omega) = \int_{-\infty}^{+\infty} e^{-i\omega t} \, \sigma_{ij}(t) \, dt \,. \tag{2.3}$$

Concerning the internal and external acoustic fluids, for the time-dependent pressure fields $p(\mathbf{x}, t)$ and $p_E(\mathbf{x}, t)$, we will also have

$$p(\mathbf{x}, \omega) = \int_{-\infty}^{+\infty} e^{-i\omega t} \, p(\mathbf{x}, t) \, dt \,, \tag{2.4}$$

$$p_E(\mathbf{x}, \omega) = \int_{-\infty}^{+\infty} e^{-i\omega t} \, p_E(\mathbf{x}, t) \, dt \,. \tag{2.5}$$

Nevertheless, for other quantities, some exceptions to this rule could be made and the Fourier transform of a function f will then be noted $\widehat{f}$,

$$\widehat{f}(\omega) = \int_{-\infty}^{+\infty} e^{-i\omega t} f(t)\, dt. \qquad (2.6)$$

2.2 STRUCTURE

In general, a complex structure is composed of a main part called the *master structure*, defined as the "primary" structure accessible to conventional modeling, and a secondary part, called the *fuzzy substructure*, related to the structural complexity. The structural complexity is made up of a great number of complex secondary subsystems such as equipment units or secondary structures attached to the master structure. These subsystems are called fuzzy substructures due to their structural complexity and because the details of them are unknown (the terminology "fuzzy" has nothing to do with the mathematical theory concerning fuzzy sets and fuzzy logic). More precisely, a fuzzy substructure is the part of the structure that is not accessible to conventional modeling because, as stated here, the number of secondary subsystems and their geometry, boundary conditions, and constitutive equations are unknown or are not accurately known. Since each fuzzy substructure is constituted by discrete or continuous weakly damped bounded structures, a fuzzy substructure is a resonant mechanical system which has a countable number of eigenfrequencies. In addition, since a fuzzy substructure is made of a large number of secondary subsystems, the modal density of such a substructure is high in the medium-frequency range of interest. Below the first eigenfrequency of a fuzzy substructure, its effects on the master structure are mainly due to added-mass effects. In this case, the morphology of the dynamic response of the master structure is not altered by the presence of fuzzy substructures. Above the first eigenfrequency of a fuzzy substructure, experimental results show that its presence induces an "apparent strong damping" in the master structure in the medium-frequency range and possibly in the low-frequency range. Therefore, this strong damping cannot be explained by the small mechanical viscous damping existing in the master structure.

In the present book, fuzzy substructures are not considered. Concerning the fuzzy structure theory, we refer the reader to Soize (1986, 1993), to chapter 15 of Ohayon and Soize (1998) for a synthesis, and to Fernandez et al. (2009) for extension of the theory to uncertain complex vibroacoustic system with fuzzy interface modeling. Consequently, the so-called "master structure" will be here called "structure."

Nevertheless, the master structure can be a complex mechanical system itself and consequently, the mathematical-mechanical process used to construct its computational model exhibits both the system-parameter uncertainties and the model uncertainties induced by modeling errors (for instance, junctions; interfaces; boundary conditions; geometrical errors due to manufacturing; introduction of kinematic reductions by the use of beam, plate, and shell theories; constitutive equations at macroscale for composite materials and, more generally, for materials with heterogeneous microstructures, such as cellular and porous materials, etc.).

It should be noted that the approach described earlier for the macroscale can be used at any scale (either for unique scale or in a multiscale framework). In addition, and depending on the available computer resources, the sophistication degree of the computational modeling of the real mechanical system can vary, but in any case, there will be always uncertainties in the computational model. This is the reason why uncertainty quantification is important for large computational vibroacoustic models.

The structure is a damped (dissipative) medium whose frequency-dependent constitutive equation is defined in Section 5.2. At equilibrium, the structure occupies the three-dimensional bounded domain Ω_S with a boundary $\partial\Omega_S$ that is constituted of three parts. The first one, Γ_E, is the coupling interface between the structure and the external acoustic fluid. The second part, Γ, is a coupling interface between the structure and the internal acoustic fluid. The last one, Γ_Z, is another part of the coupling interface between the structure and the internal acoustic fluid with acoustical properties. As previously explained, the structure is assumed to be free (free-free structure), that is, not fixed on any part of boundary $\partial \Omega_S$. The outward unit normal to $\partial \Omega_S$ is denoted as $\mathbf{n}^S = (n_1^S, n_2^S, n_3^S)$ (see Figure 2.1). In the frequency domain, the displacement field defined in Ω_S is denoted by $\mathbf{u}(\mathbf{x}, \omega) = (u_1(\mathbf{x}, \omega), u_2(\mathbf{x}, \omega), u_3(\mathbf{x}, \omega))$. A given surface force field, $\mathbf{G}(\mathbf{x}, \omega) = (G_1(\mathbf{x}, \omega), G_2(\mathbf{x}, \omega), G_3(\mathbf{x}, \omega))$, is defined on $\partial \Omega_S$ and a given body force field, $\mathbf{g}(\mathbf{x}, \omega) = (g_1(\mathbf{x}, \omega), g_2(\mathbf{x}, \omega), g_3(\mathbf{x}, \omega))$, is defined in Ω_S.

2.3 INTERNAL DISSIPATIVE ACOUSTIC FLUID

Let Ω be the internal bounded domain filled with a dissipative acoustic fluid (gas or liquid) whose equations are described in Chapter 4.

More precisely, a dissipative acoustic fluid is defined as a fluid that is homogeneous, compressible, and dissipative. In the reference configuration, the fluid is at rest. The fluid is either a gas or a liquid and gravity effects are neglected. The pressure field in Ω is denoted by $p(\mathbf{x}, \omega)$. An acoustic source density, $Q(\mathbf{x}, \omega)$, is given inside Ω and is equal to zero except in the source region where the source is active. Such a source density allows us, for example, to model a loudspeaker without modeling the loudspeaker itself as an additional vibroacoustic system.

The boundary $\partial \Omega$ of Ω is written as $\Gamma \cup \Gamma_Z$. The outward unit normal to $\partial \Omega$ is denoted as $\mathbf{n} = (n_1, n_2, n_3)$ and we have $\mathbf{n} = -\mathbf{n}^S$ on $\partial \Omega$ (see Figure 2.1).

Part Γ_Z of the boundary has acoustical properties modeled by a wall acoustic impedance $Z(\mathbf{x}, \omega)$ satisfying the hypotheses that are given in Section 4.2. This wall acoustic impedance is a modeling of the acoustic properties of the material constituting interface Γ_Z (structural material or acoustic coating material embedded in the structure or in the vicinity of its boundary).

2.4 EXTERNAL INVISCID ACOUSTIC FLUID

The structure is surrounded by an external inviscid acoustic fluid (gas or liquid). The fluid occupies the infinite three-dimensional domain Ω_E whose boundary $\partial\Omega_E$ is Γ_E. The inward unit normal to $\partial\Omega_E$ is $\mathbf{n}^S$, defined earlier (see Figure 2.1). The pressure field in Ω_E is denoted as $p_E(\mathbf{x}, \omega)$. An acoustic source density, $Q_E(\mathbf{x}, \omega)$, is given in Ω_E. Such a source density allows us, for example, to model a loudspeaker without modeling the loudspeaker itself as an additional vibroacoustic system. This acoustic source density induces a pressure field $p_{given}(\omega)$ on Γ_E (see its construction in Chapter 10). For the sake of brevity, the case of an incident plane wave is not considered and we refer the reader to Ohayon and Soize (1998) for this case.

3

EXTERNAL INVISCID ACOUSTIC
FLUID EQUATIONS

In the first section of this chapter, the equations are introduced in the frequency domain for the external inviscid acoustic fluid. The second section is devoted to the important concept of *acoustic impedance boundary operator*, which allows us to express the pressure field exerted by the external acoustic fluid on the structure in terms of the normal structural displacement field of the external fluid-structure interface.

3.1 EQUATIONS IN THE FREQUENCY DOMAIN

As presented in Section 2.4, we consider an inviscid acoustic fluid occupying the infinite domain Ω_E (see Figure 3.1), described by the acoustic pressure $p_E(\mathbf{x}, \omega)$ at point $\mathbf{x}$ of Ω_E and at circular frequency ω. Let ρ_E be the constant mass density of the external acoustic fluid at equilibrium. Let c_E be the constant speed of sound in the external acoustic fluid at equilibrium and let $k = \omega/c_E$ be the wave number at frequency ω. For a given external acoustic source density Q_E introduced in Section 2.4, and for a given normal displacement field $\mathbf{u} \cdot \mathbf{n}^S$ on Γ_E induced by the deformation of the structure and introduced in Section 2.2, the external pressure field p_E is solution of the following equations

$$\nabla^2 p_E + k^2 p_E = -i\omega Q_E \quad \text{in} \quad \Omega_E, \qquad (3.1)$$

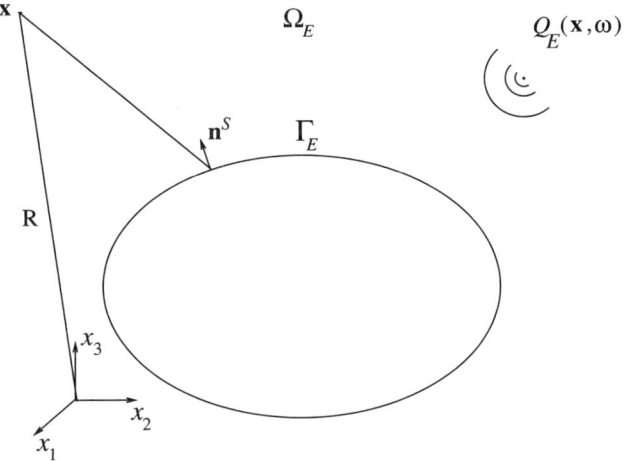

Figure 3.1 Geometry for the external acoustic fluid.

$$\frac{\partial p_E}{\partial \mathbf{n}^S} = \omega^2 \rho_E \, \mathbf{u} \cdot \mathbf{n}^S \quad \text{on} \quad \Gamma_E , \tag{3.2}$$

$$|p_E| = O(\frac{1}{R}) \, , \quad \left| \frac{\partial p_E}{\partial R} + i k \, p_E \right| = O(\frac{1}{R^2}) \, , \quad R = \|\mathbf{x}\| \to +\infty , \tag{3.3}$$

in which ∇^2 is the three-dimensional-Laplacian operator, $\partial/\partial R$ is the derivative in the radial direction, and $|z|$ denotes the modulus of the complex number z.

Equation (3.1) is the Helmholtz equation (linear wave equation formulated in the frequency domain) in terms of pressure field p_E and with a source term defined by Q_E. It is recalled that the Helmholtz equation corresponds to irrotational linearized motions of the inviscid acoustic fluid. For the derivation of this equation with source term, we refer the reader to Lighthill (1978), Pierce (1989), and Ohayon and Soize (1998).

In Eq. (3.2), the normal derivative $\partial p_E / \partial \mathbf{n}^S$ is defined as $\mathbf{n}^S \cdot \nabla p_E = n_j^S \{p_E\}_{,j}$. This equation corresponds to the continuity of the normal velocity on interface Γ_E for the external inviscid fluid and the structure, expressed in terms of the pressure field p_E and of the normal displacement $\mathbf{u} \cdot \mathbf{n}^S$. The right-hand side of Eq. (3.2) represents the opposite of the inertial forces field per unit of volume due to the acceleration field of the deformable structure on Γ_E.

Equation (3.3) is called the *outward Sommerfeld radiation condition* at infinity. In this equation, the symbol O is defined as follows. If R goes to infinity, the quantity $|p_E|$ decreases as $1/R$ and $|\partial p_E / \partial R + i k p_E|$ decreases as $1/R^2$. Therefore, the pressure field goes to zero at infinity and, for $\omega > 0$, $|p_E - \rho_E c_E v_E|$ goes to zero as $1/R^2$ when R goes to infinity, in which v_E is the velocity field in the acoustic fluid along the radial direction. These conditions show that only outward-traveling waves are considered at infinity (outward-traveling waves vanish at infinity and are not reflected) and that the radiated energy at infinity is lost.

3.2 ACOUSTIC IMPEDANCE BOUNDARY OPERATOR

As explained in Chapter 2, the exterior acoustic problem defined by Eqs. (3.1) to (3.3) will be treated in Chapter 10 by an appropriate boundary integral equation formulation. In Ohayon and Soize (1998) (including the references), it is shown that the pressure field $p_E|_{\Gamma_E}$ on interface Γ_E can be written as

$$p_E|_{\Gamma_E}(\omega) = p_{\text{given}}|_{\Gamma_E}(\omega) + i\omega \, \mathbf{Z}_{\Gamma_E}(\omega)\{\mathbf{u}(\omega) \cdot \mathbf{n}^S\}, \qquad (3.4)$$

in which $p_{\text{given}}|_{\Gamma_E}$ is the pressure field on Γ_E induced by the acoustic source density $Q_E(\mathbf{x}, \omega)$ in Ω_E and where $i\omega \mathbf{u}(\omega) \cdot \mathbf{n}^S$ is the velocity field $v(\omega)$ on Γ_E induced by the deformation of the structure.

In Eq. (3.4), $\mathbf{Z}_{\Gamma_E}(\omega)$ is a linear integral operator related to interface Γ_E. This means that, for a given field $v(\omega)$ defined on Γ_E, the quantity $\mathbf{Z}_{\Gamma_E}(\omega)\{v(\omega)\}$ (which is the image of $v(\omega)$) is a field defined on interface Γ_E. Therefore, the value $(\mathbf{Z}_{\Gamma_E}(\omega)\{v(\omega)\})(\mathbf{x})$ of $\mathbf{Z}_{\Gamma_E}(\omega)\{v(\omega)\}$ at a given point $\mathbf{x}$ in Γ_E depends on all the values of v. Operator $\mathbf{Z}_{\Gamma_E}(\omega)$ is called the *acoustic impedance boundary operator*, which is a nonlocal operator on Γ_E.

The explicit construction of $p_{\text{given}}|_{\Gamma_E}$ and of $\mathbf{Z}_{\Gamma_E}(\omega)$ is carried out in Chapter 10, which is a self-contained chapter describing the computational modeling of the exterior acoustic problem defined by Eqs. (3.1) to (3.3) by an appropriate boundary element method.

4

INTERNAL DISSIPATIVE ACOUSTIC
FLUID EQUATIONS

This chapter is devoted to the description of the equations in terms of the pressure field and the associated boundary conditions for the internal dissipative acoustic fluid of the vibroacoustic system. A wall acoustic impedance can be taken into account. In the last section, the case of a free surface for a compressible liquid is considered.

4.1 EQUATIONS IN THE FREQUENCY DOMAIN

As introduced in Section 2.3, the fluid is assumed to be homogeneous, compressible, and dissipative. In the reference configuration, the fluid is supposed to be at rest. The fluid is either a gas or a liquid and gravity effects are neglected (see Andrianarison and Ohayon, 2006, to take into account both gravity and compressibility effects for an inviscid internal fluid). Such a fluid is called a *dissipative acoustic fluid*. Generally, there are two main physical dissipations. The first one is an internal acoustic dissipation inside the cavity due to the viscosity and the thermal conduction of the fluid. These dissipation mechanisms are assumed to be small. In the model proposed, we consider only the dissipation due to the viscosity. This correction introduces an additional dissipative term in the Helmholtz equation without modifying the conservative part. The second one is the dissipation generated inside the *wall viscothermal boundary layer* of the cavity and is

24

neglected here. We then consider only the acoustic mode (irrotational motion) predominant in the volume. The vorticity and entropy modes that mainly play a role in the wall viscothermal boundary layer are not modeled. For additional details concerning dissipation in acoustic fluids, we refer the reader to Lighthill (1978); Pierce (1989); Landau and Lifchitz (1992); Beltman (1999); and Bruneau (2006).

The dissipation due to thermal conduction is neglected and the motions are assumed to be irrotational. Let ρ_0 be the constant mass density and c_0 be the constant speed of sound in the fluid at equilibrium in the reference configuration Ω. In the framework of linear dissipative acoustic fluids, from the linearized mass conservation equation, the constitutive equation for the fluid, and the linearized dynamic equation, the two following equations can be deduced in the frequency domain (see Ohayon and Soize, 1998),

$$i\omega\, p = -\rho_0\, c_0^2\, \nabla \cdot \mathbf{v} + c_0^2\, Q\,, \tag{4.1}$$

$$i\omega\, \rho_0\, \mathbf{v} + \nabla p = \tau c_0^2 \nabla Q - i\omega\, \tau \nabla p\,, \tag{4.2}$$

in which $p(\mathbf{x}, \omega)$ is the pressure field in Ω, where $\mathbf{v}(\mathbf{x}, \omega)$ is the velocity field in Ω, and where τ is given by

$$\tau = \frac{1}{\rho_0 c_0^2}\left(\frac{4}{3}\eta + \zeta\right) > 0\,. \tag{4.3}$$

The constant η is the dynamic viscosity, $\nu = \eta/\rho_0$ is the kinematic viscosity, and ζ is the second viscosity. Eliminating $\mathbf{v}$ between Eqs. (4.1) and (4.2), then dividing by ρ_0, yield the Helmholtz equation with a dissipative term and a right-hand side member induced by the given acoustic source,

$$-\frac{\omega^2}{\rho_0 c_0^2}\, p - i\omega\frac{\tau}{\rho_0}\nabla^2 p - \frac{1}{\rho_0}\,\nabla^2 p = \frac{1}{\rho_0}(i\omega Q - \tau c_0^2\nabla^2 Q) \quad \text{in} \quad \Omega\,. \tag{4.4}$$

The coefficient τ characterizes the dissipation in the internal acoustic fluid. Let us remark that taking $\tau = 0$ and $Q = 0$ in Eq. (4.4) yields

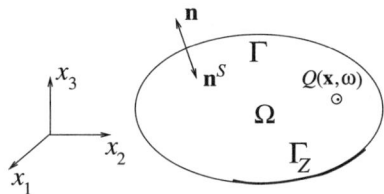

Figure 4.1   Geometry for the internal acoustic fluid.

the Helmholtz equation for wave propagation in inviscid acoustic fluid.

4.2 BOUNDARY CONDITIONS IN THE FREQUENCY DOMAIN

We now define the boundary conditions on Γ and Γ_Z (see Figure 4.1).

(i) Boundary condition on Γ. Using Eq. (4.2) and the continuity on Γ of the normal acoustic fluid velocity $\mathbf{v} \cdot \mathbf{n}$ with the normal structural velocity $i\omega\, \mathbf{u} \cdot \mathbf{n}$ yields the following boundary condition:

$$(1 + i\omega\, \tau)\, \frac{\partial p}{\partial \mathbf{n}} = \omega^2\, \rho_0\, \mathbf{u} \cdot \mathbf{n} + \tau\, c_0^2\, \frac{\partial Q}{\partial \mathbf{n}} \quad \text{on} \quad \Gamma. \qquad (4.5)$$

Taking $\tau = 0$ and $Q = 0$ in Eq. (4.5) yields the boundary condition corresponding to the continuity of the normal velocity on interface Γ for the inviscid fluid and the structure, expressed in terms of the pressure field p and the normal displacement $\mathbf{u} \cdot \mathbf{n}$.

(ii) Boundary condition with wall acoustic impedance. The part Γ_Z of the boundary $\partial\Omega$ has acoustic properties modeled by a *wall acoustic impedance* $Z(\mathbf{x}, \omega)$ defined for $\mathbf{x}$ in Γ_Z, with complex values. The wall impedance boundary condition on Γ_Z is written as

$$p(\mathbf{x}, \omega) = Z(\mathbf{x}, \omega)\, \{\mathbf{v}(\mathbf{x}, \omega) \cdot \mathbf{n} - i\omega\, \mathbf{u}(\mathbf{x}, \omega) \cdot \mathbf{n}\}. \qquad (4.6)$$

Wall acoustic impedance $Z(\mathbf{x}, \omega)$ must satisfy appropriate conditions in order to ensure that the problem is correctly stated (see Ohayon and Soize, 1998, for a general formulation, and see Deü et al., 2008,

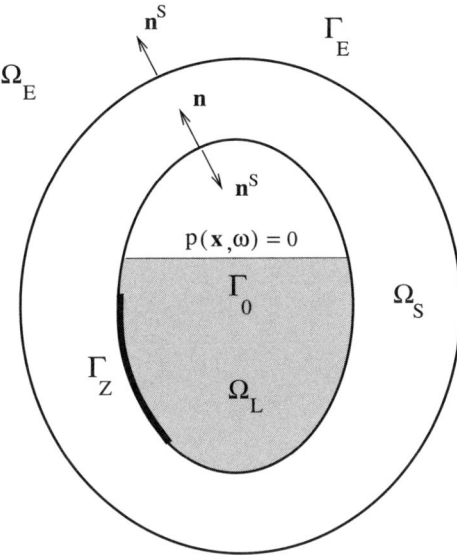

Figure 4.2 Configuration of the structural-acoustic system for a liquid with free surface.

for a simplified model of the Voigt type with an internal inviscid fluid). Taking the normal component of Eq. (4.2) on Γ_Z with respect to $\mathbf{n}$, and substituting in it the value of $\mathbf{v} \cdot \mathbf{n}$ deduced from Eq. (4.6), yields the following boundary condition on Γ_Z, with wall acoustic impedance:

$$(1 + i\omega \tau) \frac{\partial p}{\partial \mathbf{n}} = \omega^2 \rho_0 \mathbf{u} \cdot \mathbf{n} - i\omega\rho_0 \frac{p}{Z} + \tau c_0^2 \frac{\partial Q}{\partial \mathbf{n}} \quad \text{on} \quad \Gamma_Z . \quad (4.7)$$

4.3 CASE OF A FREE SURFACE FOR A LIQUID

Cavity Ω is partially filled with a liquid (dissipative acoustic fluid) occupying the domain Ω_L. It is assumed that the complementary part $\Omega \backslash \Omega_L$ is a vacuum domain. The boundary $\partial\Omega_L$ of Ω_L is constituted of three boundaries, Γ_Z, Γ_0 corresponding to the free surface of the liquid, and a part Γ_L of Γ (see Figure 4.2). The Neumann boundary conditions are given by Eq. (4.5) on Γ_L and by Eq. (4.7) on Γ_Z. Neglecting the dynamic effects in the presence of gravity (of course, the existence of a static equilibrium with the presence of a free surface is due to gravity), the following Dirichlet boundary condition is written on the free surface,

$$p = 0 \quad \text{on} \quad \Gamma_0. \tag{4.8}$$

5

STRUCTURE EQUATIONS

In this chapter, the equations and the associated boundary conditions are given for the structure in the frequency domain. The frequency-dependent constitutive equation is detailed for the dissipative structure. In the medium-frequency range, it is important to take into account the frequency dependence of the coefficients of the constitutive equation in order to correctly model the damping and the dissipative effects. A complex heterogeneous structure is made up of many parts relevant to various types of damping modeling associated with different materials, such as viscoelastic materials (modeled in Section 5.2.1). It can also be made up of dissipative elements such as joints, attachments, connections, and other complex interfaces. In general, the mechanical behavior of such complex interfaces requires the use of dissipative constitutive equation for modeling damping effects in low- and medium-frequency domains (modeled in Section 5.2.2). These two cases are relevant to the same type of constitutive equation whose frequency dependent coefficients are differently constructed.

5.1 STRUCTURE EQUATIONS
IN THE FREQUENCY DOMAIN

In the frequency domain, the linearized dynamic equation of the structure around the static equilibrium state at rest, occupying domain Ω_S, is written as

$$-\omega^2 \rho_S \mathbf{u} - \mathrm{div}\, \sigma = \mathbf{g} \quad \text{in} \quad \Omega_S, \tag{5.1}$$

in which $\rho_S(\mathbf{x})$ is the mass density of the structure; $\mathbf{u}(\mathbf{x}, \omega) = (u_1(\mathbf{x}, \omega),$ $u_2(\mathbf{x}, \omega),\ u_3(\mathbf{x}, \omega))$ is the displacement field; $\mathbf{g}(\mathbf{x}, \omega) = (g_1(\mathbf{x}, \omega),$ $g_2(\mathbf{x}, \omega),\ g_3(\mathbf{x}, \omega))$ is the given body force field defined in Ω_S; and where the divergence of the stress tensor $\sigma = \{\sigma_{ij}\}_{i,j=1,2,3}$ is such that $\{\mathrm{div}\,\sigma\}_i = \sigma_{ij,j}$. The frequency-dependent constitutive equation is defined in the following and will be analyzed in Section 5.2.1 for linear viscoelastic material and in Section 5.2.2 for linear elastic material with damping effects. The symmetric stress tensor σ (i.e. $\sigma_{ij} = \sigma_{ji}$) is then written as

$$\sigma_{ij} = (a_{ijkh}(\mathbf{x}, \omega) + i\omega\, b_{ijkh}(\mathbf{x}, \omega))\, \varepsilon_{kh}, \tag{5.2}$$

in which the symmetric strain tensor $\varepsilon = \{\varepsilon_{kh}\}_{k,h=1,2,3}$ (i.e. $\varepsilon_{kh} = \varepsilon_{hk}$) is such that

$$\varepsilon_{kh} = \frac{1}{2}(u_{k,h}(\mathbf{x}, \omega) + u_{h,k}(\mathbf{x}, \omega)). \tag{5.3}$$

The real fourth-order tensor $\{a_{ijkh}(\mathbf{x}, \omega)\}_{i,j,k,h=1,2,3}$ and the real fourth-order tensor $\{b_{ijkh}(\mathbf{x}, \omega)\}_{i,j,k,h=1,2,3}$ depend on ω (detailed in Section 5.2) and depend on $\mathbf{x}$ because the material is assumed to be heterogeneous.

The boundary conditions concerning the forces acting on the structure (see Figure 5.1) are defined in the following. On the fluid-structure external interface Γ_E, the surface force fields acting on the structure are the given surface force field $\mathbf{G}(\mathbf{x}, \omega) = (G_1(\mathbf{x}, \omega), G_2(\mathbf{x}, \omega), G_3(\mathbf{x}, \omega))$ and the pressure field $p_E|_{\Gamma_E}(\mathbf{x}, \omega)$ exerted by the external acoustic fluid. We then have

$$\sigma \mathbf{n}^S = \mathbf{G} - p_E|_{\Gamma_E}\, \mathbf{n}^S \quad \text{on} \quad \Gamma_E, \tag{5.4}$$

in which $\{\sigma \mathbf{n}^S\}_i = \sigma_{ij} n_j^S$ and where $p_E|_{\Gamma_E}(\mathbf{x}, \omega)$ is given (see Eq. (3.4)) by

$$p_E|_{\Gamma_E} = p_{\text{given}}|_{\Gamma_E} + i\omega\, \mathbf{Z}_{\Gamma_E}(\omega)\{\mathbf{u} \cdot \mathbf{n}^S\} \quad \text{on} \quad \Gamma_E. \tag{5.5}$$

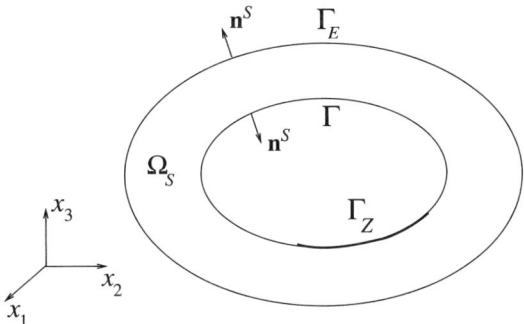

Figure 5.1 Configuration for the structure.

On the fluid-structure internal interface $\Gamma \cup \Gamma_Z$, the surface force fields acting on the structure are the given surface force field $\mathbf{G}(\mathbf{x}, \omega)$ and the internal acoustic pressure field $p(\mathbf{x}, \omega)$ exerted by the internal acoustic fluid introduced in Chapter 4. We then have

$$\sigma \, \mathbf{n}^S = \mathbf{G} - p \, \mathbf{n}^S \quad \text{on} \quad \Gamma \cup \Gamma_Z . \tag{5.6}$$

5.2 FREQUENCY-DEPENDENT LINEAR CONSTITUTIVE EQUATIONS

Two cases of frequency-dependent linear constitutive equations are considered in order to describe all the various types of mechanical behaviors encountered in a complex structure.

The first one is relevant to the framework of the general linear viscoelasticity theory for describing the constitutive equation of viscoelastic materials and, therefore, the frequency-dependent coefficients are constructed in this framework, which guarantees the causality physical property. This constitutive equation will be referred as the *linear viscoelastic constitutive equation*.

The second one allows different types of mechanical damping to be modeled using the same type of constitutive equation. The frequency-dependent coefficients will not be constructed in the framework of the linear viscoelasticity theory but will be constructed in such a way that the causality physical property will still be satisfied. This constitutive equation is referred to as the *linear dissipative constitutive equation for modeling damping effects*.

5.2.1 Linear Viscoelastic Constitutive Equation

The general theory of linear viscoelasticity is used (see Bland, 1960; Fung, 1968; Truesdell, 1973). With respect to the presentation detailed in Ohayon and Soize (1998), we present here a summary of those results with additional developments.

In this section, $\mathbf{x}$ is fixed in Ω_S and will be omitted in all the quantities. The Latin indices, such as i, j, k, and h, take the values 1, 2, and 3. The convention for summations over repeated Latin indices is used.

Constitutive Equation in the Time Domain
The components $\sigma_{ij}(t)$ of the stress tensor $\sigma(t)$ are written as

$$\sigma_{ij}(t) = \mathcal{G}_{ijkh}(0)\,\varepsilon_{kh}(t) + \int_0^{+\infty} \dot{\mathcal{G}}_{ijkh}(\tau)\varepsilon_{kh}(t-\tau)d\tau\,, \qquad (5.7)$$

where $\sigma_{ij}(t) = 0$ and $\varepsilon_{kh}(t) = 0$ for $t \leq 0$. The real functions $\mathcal{G}_{ijkh}(t)$ are called the *relaxation functions*. Function $\mathcal{G}_{ijkh}(t)$ (and thus $\dot{\mathcal{G}}_{ijkh}(t)$) has the property of symmetry, $\mathcal{G}_{ijkh}(t) = \mathcal{G}_{jikh}(t) = \mathcal{G}_{ijhk}(t) = \mathcal{G}_{khij}(t)$. The value $\mathcal{G}_{ijkh}(0)$ at time $t = 0$, which is called the *initial elasticity tensor*, is positive definite. This means that, for all real symmetric tensors X_{ij} such that $X_{ij}X_{ij}$ is strictly positive, the real quantity $\mathcal{G}_{ijkh}(0)\,X_{kh}X_{ij}$ is strictly positive. The relaxation functions are defined on $[0, +\infty[$, they are differentiable with respect to t on $]0, +\infty[$, and their derivatives are denoted as $\dot{\mathcal{G}}_{ijkh}(t)$ and are assumed to be integrable on $[0, +\infty[$, which means that $\int_0^{+\infty} |\dot{\mathcal{G}}_{ijkh}(t)|\, dt$ is finite. Functions $\mathcal{G}_{ijkh}(t)$ can be written as

$$\mathcal{G}_{ijkh}(t) = \mathcal{G}_{ijkh}(0) + \int_0^t \dot{\mathcal{G}}_{ijkh}(\tau)\, d\tau\,. \qquad (5.8)$$

Therefore, the limit of $\mathcal{G}_{ijkh}(t)$ as t tends to $+\infty$, denoted as $\mathcal{G}_{ijkh}(\infty)$, is finite and is given by

$$\mathcal{G}_{ijkh}(\infty) = \mathcal{G}_{ijkh}(0) + \int_0^{+\infty} \dot{\mathcal{G}}_{ijkh}(\tau)\, d\tau. \tag{5.9}$$

The fourth-order tensor $\mathcal{G}_{ijkh}(\infty)$, called the *equilibrium modulus tensor*, is symmetric, is positive definite, and corresponds to the elasticity coefficients of the elastic material for a static deformation. In effect, it can be shown that the static equilibrium state is obtained for t tends to infinity. Let us then introduce the real functions $g_{ijkh}(t)$, such that

$$g_{ijkh}(t) = 0 \quad \text{if} \quad t < 0, \tag{5.10}$$

$$g_{ijkh}(t) = \dot{\mathcal{G}}_{ijkh}(t) \quad \text{if} \quad t \geq 0. \tag{5.11}$$

Since $g_{ijkh}(t) = 0$ for $t < 0$, it can be deduced that $g_{ijkh}(t)$ is a causal function. The terminology "causal function" comes from the linear filtering theory (see, for instance, Papoulis, 1977). In effect, let $g(t)$ be the impulse response function of a convolution linear filter whose input is any function $e(t)$. Then, at time t, the filter output $o(t)$ is written as $o(t) = \int_{\mathbb{R}} g(\tau)\, e(t - \tau)\, d\tau$. If function $g(t)$ is such that $g(t) = 0$ for $t < 0$, then it can be seen that $o(t)$ depends only on the past $\{e(t'), t' \leq t\}$ and not on the future $\{e(t'), t' > t\}$, because we have $o(t) = \int_0^{+\infty} g(\tau)\, e(t - \tau)\, d\tau = \int_{-\infty}^t g(t - \tau)\, e(\tau)\, d\tau$.

Using Eq. (5.11), Eq. (5.7) can be rewritten as

$$\sigma_{ij}(t) = \mathcal{G}_{ijkh}(0)\, \varepsilon_{kh}(t) + \int_{-\infty}^{+\infty} g_{ijkh}(\tau)\varepsilon_{kh}(t - \tau)\, d\tau. \tag{5.12}$$

It should be noted that Eq. (5.12) corresponds, in the time domain, to the general formulation of the linear theory of viscoelasticity. The usual approach, which consists in modeling the constitutive equation in the time domain by a linear differential equation in $\sigma(t)$ and $\varepsilon(t)$ (see, for instance, Truesdell, 1973; Dautray and Lions, 1992), corresponds to a particular case which is an approximation of Eq. (5.12). An alternative approximation of Eq. (5.12) consists in representing the integral operator by a differential operator acting on additional hidden variables. This type of approximation can efficiently be described using fractional derivative operators (see, for instance, Bagley and Torvik, 1983; Deü and Matignon, 2010).

Linear Viscoelastic Constitutive Equation in the Frequency Domain
Taking the Fourier transform of the two side members of Eq. (5.12), it is shown as follows that the general constitutive equation in the frequency range can be written as

$$\sigma_{ij}(\omega) = (a_{ijkh}(\omega) + i\omega\, b_{ijkh}(\omega))\, \varepsilon_{kh}(\omega)\,. \qquad (5.13)$$

In effect, since $g_{ijkh}(t)$ is an integrable function on $]-\infty, +\infty[$, its Fourier transform $\widehat{g}_{ijkh}(\omega)$ is defined by

$$\begin{aligned}
\widehat{g}_{ijkh}(\omega) &= \int_{-\infty}^{+\infty} e^{-i\omega t}\, g_{ijkh}(t)\, dt \\
&= \int_{0}^{+\infty} e^{-i\omega t}\, \dot{\mathcal{G}}_{ijkh}(t)\, dt\,.
\end{aligned} \qquad (5.14)$$

Due to the integrability of function $g_{ijkh}(t)$, function $\widehat{g}_{ijkh}(\omega)$ is a complex function, which is continuous on $]-\infty, +\infty[$ and is such that

$$\lim_{|\omega| \to +\infty} |\widehat{g}_{ijkh}(\omega)| = 0\,. \qquad (5.15)$$

Then, introducing the real part $\widehat{g}_{ijkh}^{R}(\omega) = \Re e\{\widehat{g}_{ijkh}(\omega)\}$ and the imaginary part $\widehat{g}_{ijkh}^{I}(\omega) = \Im m\{\widehat{g}_{ijkh}(\omega)\}$, it can be seen that

$$\widehat{g}_{ijkh}^{R}(-\omega) = \widehat{g}_{ijkh}^{R}(\omega)\quad, \quad \widehat{g}_{ijkh}^{I}(-\omega) = -\widehat{g}_{ijkh}^{I}(\omega)\,. \qquad (5.16)$$

We then have

$$\widehat{g}_{ijkh}^{I}(0) = 0\,. \qquad (5.17)$$

Since the Fourier transform of Eq. (5.12) is written as

$$\sigma_{ij}(\omega) = (\mathcal{G}_{ijkh}(0) + \widehat{g}_{ijkh}^{R}(\omega) + i\,\widehat{g}_{ijkh}^{I}(\omega))\, \varepsilon_{kh}(\omega)\,, \qquad (5.18)$$

from Eq. (5.13), it can be deduced that

$$a_{ijkh}(\omega) = \mathcal{G}_{ijkh}(0) + \widehat{g}^R_{ijkh}(\omega), \tag{5.19}$$

$$\omega\, b_{ijkh}(\omega) = \widehat{g}^I_{ijkh}(\omega). \tag{5.20}$$

From Eqs. (5.15), (5.19), and (5.20), it can be deduced that

$$\lim_{|\omega|\to+\infty} a_{ijkh}(\omega) = \mathcal{G}_{ijkh}(0), \tag{5.21}$$

$$\lim_{|\omega|\to+\infty} \omega\, b_{ijkh}(\omega) = 0. \tag{5.22}$$

From Eqs. (5.13), (5.21), and (5.22), it can be deduced that

$$\sigma_{ij}(\infty) = \mathcal{G}_{ijkh}(0)\, \varepsilon_{kh}(\infty). \tag{5.23}$$

Equation (5.23) shows that viscoelastic materials behave elastically at high frequencies with elasticity coefficients defined by the *initial elasticity tensor* $\mathcal{G}_{ijkh}(0)$, which differs from the *equilibrium modulus tensor* $\mathcal{G}_{ijkh}(\infty)$. Using Eqs. (5.9) and (5.17), this tensor is written as

$$\mathcal{G}_{ijkh}(\infty) = \mathcal{G}_{ijkh}(0) + \widehat{g}^R_{ijkh}(0). \tag{5.24}$$

The viscoelastic material is dissipative, which means that, for all $t \geq 0$, we have $\int_0^t \sigma_{ij}(\tau)\dot{\varepsilon}_{ij}(\tau)\, d\tau > 0$. Using this inequality, for sufficiently smooth ε_{ij} with $\varepsilon_{ij}(t)_{t=0} = 0$, Gurtin and Herrera (1965) proved that initial elasticity tensor $\mathcal{G}_{ijkh}(0)$ and equilibrium modulus tensor $\mathcal{G}_{ijkh}(\infty)$ are positive-definite symmetric fourth-order tensors. The tensor $\mathcal{G}_{ijkh}(\infty)$ corresponds to the elasticity coefficients of a linear elastic material for a static deformation process. More specifically, using Eq. (5.13) for $\omega = 0$, (5.17) and (5.20) yield

$$\sigma_{ijkh}(0) = a_{ijkh}(0)\, \varepsilon_{ijkh}(0). \tag{5.25}$$

in which $\sigma_{ijkh}(0) = \{\sigma_{ijkh}(\omega)\}_{\omega=0}$ and $\varepsilon_{ijkh}(0) = \{\varepsilon_{ijkh}(\omega)\}_{\omega=0}$, and where

$$a_{ijkh}(0) = \mathcal{G}_{ijkh}(0) + \widehat{g}^R_{ijkh}(0) = \mathcal{G}_{ijkh}(\infty). \tag{5.26}$$

The reader should be aware of the fact that the constitutive equation of an elastic material in a static deformation process is defined by equilibrium modulus tensor $\mathcal{G}_{ijkh}(\infty)$ and not by the initial elasticity tensor $\mathcal{G}_{ijkh}(0)$. Referring to Coleman (1964) and Truesdell (1973), it has been proven that $\mathcal{G}_{ijkh}(0) - \mathcal{G}_{ijkh}(\infty)$ is a positive-definite tensor and, consequently, $\widehat{g}^{R}_{ijkh}(0) = \mathcal{G}_{ijkh}(\infty) - \mathcal{G}_{ijkh}(0)$ is a negative-definite tensor.

Even property for viscoelastic coefficients. From Eqs. (5.19), (5.20), and (5.16), it can be deduced that tensors $a_{ijkh}(\omega)$ and $b_{ijkh}(\omega)$ are even functions,

$$a_{ijkh}(-\omega) = a_{ijkh}(\omega) \quad , \quad b_{ijkh}(-\omega) = b_{ijkh}(\omega) . \qquad (5.27)$$

Symmetry and positiveness properties. Due to the symmetry properties of tensors $\mathcal{G}_{ijkh}(t)$, it can directly be deduced that tensors $a_{ijkh}(\omega)$ and $b_{ijkh}(\omega)$ must satisfy the symmetry properties

$$a_{ijkh}(\omega) = a_{jikh}(\omega) = a_{ijhk}(\omega) = a_{khij}(\omega) , \qquad (5.28)$$

$$b_{ijkh}(\omega) = b_{jikh}(\omega) = b_{ijhk}(\omega) = b_{khij}(\omega) . \qquad (5.29)$$

In addition, the following positive-definiteness properties can be shown. For all second-order real symmetric tensors X_{ij},

$$a_{ijkh}(\omega) \, X_{kh} \, X_{ij} \geq c_a(\omega) \, X_{ij} \, X_{ij} , \qquad (5.30)$$

$$b_{ijkh}(\omega) \, X_{kh} \, X_{ij} \geq c_b(\omega) \, X_{ij} \, X_{ij} , \qquad (5.31)$$

in which the positive constants $c_a(\omega)$ and $c_b(\omega)$ are such that $c_a(\omega) \geq c_0 > 0$ and $c_b(\omega) \geq c_0 > 0$, where c_0 is a positive real constant independent of ω.

Compatibility equations between $\widehat{g}^{R}_{ijkh}(\omega)$ and $\widehat{g}^{I}_{ijkh}(\omega)$ induced by the causality property of $g_{ijkh}(t)$. Since $g_{ijkh}(t)$ is a causal function, the real part $\widehat{g}^{R}_{ijkh}(\omega)$ and the imaginary part $\widehat{g}^{I}_{ijkh}(\omega)$ of its Fourier transform $\widehat{g}_{ijkh}(\omega)$ are related by the following relations involving the

Hilbert transform (see Papoulis, 1977; Hahn, 1996; Pandey, 1996; King, 2009; Feldman, 2011),

$$\widehat{g}^R_{ijkh}(\omega) = \frac{1}{\pi} \text{p.v} \int_{-\infty}^{+\infty} \frac{\widehat{g}^I_{ijkh}(\omega')}{\omega - \omega'} \, d\omega', \qquad (5.32)$$

$$\widehat{g}^I_{ijkh}(\omega) = -\frac{1}{\pi} \text{p.v} \int_{-\infty}^{+\infty} \frac{\widehat{g}^R_{ijkh}(\omega')}{\omega - \omega'} \, d\omega', \qquad (5.33)$$

in which p.v denotes the Cauchy principal value. If $y \mapsto h(y)$ is a locally integrable function on the real line except in a singular point $y = 0$, then the p.v is defined as

$$\text{p.v} \int_{-\infty}^{+\infty} h(y) \, dy = \lim_{\ell \to +\infty, \eta \to 0^+} \{ \int_{-\ell}^{-\eta} h(y) \, dy + \int_{\eta}^{\ell} h(y) \, dy \}. \qquad (5.34)$$

The relations defined by Eqs. (5.32) and (5.33) are also called the Kramers and Kronig relations for function $g_{ijkh}(t)$ (see Kronig, 1926; Kramers, 1927). It should be noted that Eqs. (5.32) and (5.33) are not independent equations. Equation (5.32) allows $\widehat{g}^R_{ijkh}$ to be calculated if $\widehat{g}^I_{ijkh}$ is given, while Eq. (5.33) allows $\widehat{g}^I_{ijkh}$ to be calculated if $\widehat{g}^R_{ijkh}$ is given. Consequently, functions $a_{ijkh}(\omega)$ and $b_{ijkh}(\omega)$ cannot be arbitrarily chosen.

Relationship between a_{ijkh} and b_{ijkh}. We recall that $a_{ijkh}(0)$ is the equilibrium modulus tensor (see Eq. (5.26)) which is the elastic tensor and which is denoted by a^{elas}_{ijkh},

$$a^{\text{elas}}_{ijkh} = a_{ijkh}(0) . \qquad (5.35)$$

The relationship between a_{ijkh} and b_{ijkh} is written as,

$$a_{ijkh}(\omega) = a^{\text{elas}}_{ijkh} + \frac{\omega}{\pi} \text{p.v} \int_{-\infty}^{+\infty} \frac{b_{ijkh}(\omega')}{\omega - \omega'} \, d\omega' . \qquad (5.36)$$

Proof. Equations (5.20) and (5.32) yield

$$\widehat{g}^R_{ijkh}(\omega) = \frac{1}{\pi} \text{p.v} \int_{-\infty}^{+\infty} \frac{\omega' \, b_{ijkh}(\omega')}{\omega - \omega'} \, d\omega' . \qquad (5.37)$$

Equations (5.19), and (5.26) yield

$$a_{ijkh}(\omega) = a_{ijkh}(0) + \widehat{g}^R_{ijkh}(\omega) - \widehat{g}^R_{ijkh}(0).$$ (5.38)

Finally, Eq.(5.36) is obtained by substituting Eq. (5.37) in Eq. (5.38) and using Eq. (5.35).

Construction of the Linear Viscoelastic Constitutive Equation in the Frequency Domain

Two cases are considered.

(i) *Particular case.* A family of linear viscoelastic constitutive equations can be constructed in the time domain using linear differential equations in $\sigma(t)$ and $\varepsilon(t)$. The associated frequency-dependent coefficients $a_{ijkh}(\omega)$ and $b_{ijkh}(\omega)$ automatically verify Eq. (5.36). In this framework, some examples for $a_{ijkh}(\omega)$ and $b_{ijkh}(\omega)$ can be found in the literature (see, for instance, Bland, 1960; Truesdell, 1973; Bagley and Torvik, 1983; Golla and Hughes, 1985; Lesieutre and Mingori, 1990; Dautray and Lions, 1992; Mc Tavish and Hughes, 1993; Dovstam, 1995; Ohayon and Soize, 1998; Lesieutre, 2010).

(ii) *General case.* In the general case for which $a_{ijkh}(\omega)$ and $b_{ijkh}(\omega)$ are not derived from such an algebraic representation but correspond to a general integral operator in the time domain (for instance, constructed using experimental curves), a rigorous method of construction is proposed in the following to satisfy the causality principle.

For the general case, it is assumed that a part Ω_{visco} of the structure Ω_S is made of material modeled in the framework of the linear viscoelasticity theory (see the following) while the complementary part Ω_{damp} will be modeled with a linear dissipative constitutive equation for modeling damping effects (and detailed in Section 5.2.2). We then have $\Omega_S = \Omega_{\text{visco}} \cup \Omega_{\text{damp}}$.

For the practical construction of the constitutive equation related to Ω_{visco}, it is assumed that functions $\omega \mapsto b_{ijkh}(\mathbf{x}, \omega)$ for $\omega \geq 0$ and equilibrium modulus tensor $a^{\text{elas}}_{ijkh}(\mathbf{x})$ (which is the symmetric and positive-definite elastic tensor) are given. For real ω and for $\mathbf{x}$ belonging to Ω_{visco}, functions $\omega \mapsto a_{ijkh}(\mathbf{x}, \omega)$ can then be constructed.

- The given functions $\omega \mapsto b_{ijkh}(\mathbf{x}, \omega)$ cannot be arbitrary functions but must satisfy some hypotheses to ensure the coherence of the viscoelastic model:

  (1) For all fixed $\mathbf{x}$ and ω, the tensor $\{b_{ijkh}(\mathbf{x}, \omega)\}_{ijkh}$ must be symmetric and positive definite.

  (2) For $\omega \to +\infty$, Eq. (5.22) must hold, which means that functions $b_{ijkh}(\mathbf{x}, \omega)$ decrease at infinity at least in $\omega^{-\alpha}$ with $\alpha > 1$.

  (3) Functions $\omega \mapsto b_{ijkh}(\mathbf{x}, \omega)$ that satisfy (1) and (2) are then extended to $\omega < 0$ using the even property defined by Eq. (5.27).

- For all fixed $\mathbf{x}$ and ω, the tensor $\{a_{ijkh}(\mathbf{x}, \omega)\}_{ijkh}$ must be symmetric and positive definite. For all $\omega \geq 0$, functions $\omega \mapsto a_{ijkh}(\mathbf{x}, \omega)$ are then constructed using the following equation (see Eq. (5.36)),

$$a_{ijkh}(\mathbf{x}, \omega) = a_{ijkh}^{\text{elas}}(\mathbf{x}) + \frac{\omega}{\pi} \, \text{p.v} \int_{-\infty}^{+\infty} \frac{b_{ijkh}(\mathbf{x}, \omega')}{\omega - \omega'} \, d\omega', \quad (5.39)$$

and are extended to $\omega < 0$ using the even property.

- As seen earlier, for all fixed $\mathbf{x}$ and ω, symmetric tensor $\{a_{ijkh}(\mathbf{x}, \omega)\}_{ijkh}$ must be positive definite. This property must then be checked at the end of the construction and if it is not satisfied, functions $\omega \mapsto b_{ijkh}(\mathbf{x}, \omega)$ must be modified. In Soize and Poloskov (2012), it has been shown that the following sufficient condition allows this property to be satisfied: if functions $\omega \mapsto b_{ijkh}(\mathbf{x}, \omega)$ are decreasing function for $\omega \geq 0$, then the property is verified.

5.2.2 Linear Dissipative Constitutive Equation for Modeling Damping Effects

This section deals with the linear dissipative constitutive equation for modeling damping effects in the part Ω_{damp} of the structure Ω_S. Several models of dissipative constitutive equation (with frequency dependent coefficients) corresponding to an elastic material are considered for which the mechanical damping is arbitrarily introduced in order to represent damping effects. The first model presented is the constitutive equation for an elastic material with a linear viscous damping term. The second one is a constitutive equation for an elastic material with a parameterized family of damping models depending on frequency. The construction proposed for these two models is such that the causality principle will be verified and, consequently, the fourth-order tensor $a_{ijkh}(\omega)$ will depend on ω although the elastic tensor a_{ijkh}^{elas} of the elastic material is independent of ω.

(i) *Constitutive equation for an elastic material with a linear viscous damping term.* In this case, the constitutive equation is given by Eq. (5.2), in which

$$a_{ijkh}(\mathbf{x}, \omega) = a_{ijkh}^{\text{elas}}(\mathbf{x}) \quad , \quad b_{ijkh}(\mathbf{x}, \omega) = b_{ijkh}(\mathbf{x}) , \qquad (5.40)$$

in which the tensors $a_{ijkh}^{\text{elas}}(\mathbf{x})$ and $b_{ijkh}(\mathbf{x})$ are symmetric, positive-definite, and independent of ω.

(ii) *Constitutive equation for an elastic material with a parameterized family of damping models depending on frequency.* The constitutive equation is then defined as an elastic material with a parameterized family of damping models depending on frequency and written as

$$b_{ijkh}(\mathbf{x}, \omega) = \chi(\omega) \, a_{ijkh}^{\text{elas}}(\mathbf{x}) , \qquad (5.41)$$

in which the tensor $a_{ijkh}^{\text{elas}}(\mathbf{x})$ is symmetric positive definite and where $\chi(\omega)$ is a positive-valued real function in ω, which must satisfy the following properties deduced from Section 5.2.1 (ii):

(1) For $\omega \to +\infty$, function $\chi(\omega)$ must decrease at infinity at least in $\omega^{-\alpha}$ in which $\alpha > 1$.
(2) Function χ is even, $\chi(-\omega) = \chi(\omega)$.

From Eq. (5.39), it can be deduced that, for all fixed $\mathbf{x}$ and for all $\omega \geq 0$, the symmetric positive definite tensor must be constructed by the following equation:

$$a_{ijkh}(\mathbf{x}, \omega) = \left\{1 + \frac{\omega}{\pi} \text{ p.v} \int_{-\infty}^{+\infty} \frac{\chi(\omega')}{\omega - \omega'} \, d\omega'\right\} a_{ijkh}^{\text{elas}}(\mathbf{x}), \quad (5.42)$$

in order to ensure the causality property for the constitutive equation. For $\omega < 0$, the values of $a_{ijkh}(\mathbf{x}, \omega)$ are obtained using the even property in ω of functions $a_{ijkh}(\mathbf{x}, \omega)$. Finally, function $\chi(\omega)$ must be such that, for all fixed $\mathbf{x}$ and ω, symmetric tensor $\{a_{ijkh}(\mathbf{x}, \omega)\}_{ijkh}$ is positive definite. As previously explained, this property will be satisfied if function $\chi(\omega)$ is a decreasing function for $\omega \geq 0$.

6

VIBROACOUSTIC BOUNDARY
VALUE PROBLEM

In this chapter, the vibroacoustic boundary value problem is presented in terms of the external acoustic pressure field p_E, the structural displacement field $\mathbf{u}$ and the internal acoustic pressure field p. Then, using the acoustic impedance boundary operator $\mathbf{Z}_{\Gamma_E}(\omega)$ introduced in Chapter 3 and constructed in Chapter 10, the field p_E is eliminated in order to construct the equivalent vibroacoustic boundary value problem in terms of $(\mathbf{u}, p)$, which constitutes the basic formulation of the vibroacoustic boundary value problem that will be used for the computational issues. Finally, the cases of internal and external acoustic liquid with a free surface are presented.

6.1 VIBROACOUSTIC BOUNDARY VALUE PROBLEM
IN $(p_E, \mathbf{u}, p)$

The vibroacoustic system is submitted to a surface force field $\mathbf{G}(\omega)$ applied on the structure, to a body force field $\mathbf{g}(\omega)$ applied in the structure, to an external acoustic source density $Q_E(\omega)$ inside the external acoustic fluid, and to an internal acoustic source density $Q(\omega)$ inside the internal acoustic fluid. We are interested in studying the linear vibrations of the vibroacoustic system around a static equilibrium which is considered as a natural state at rest (the external structural forces and the forces induced by the acoustic sources are assumed

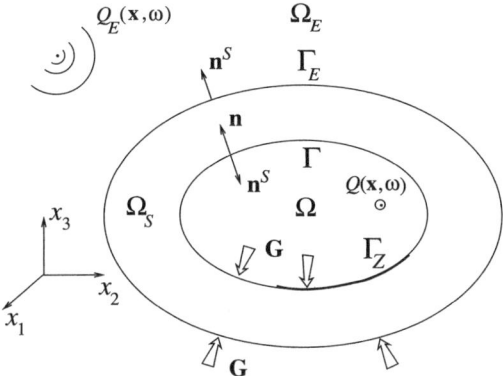

Figure 6.1 Configuration of the vibroacoustic system.

to be in equilibrium). The configuration of the vibroacoustic system is defined in Figure 6.1. For this vibroacoustic system, the physical and mechanical hypotheses are defined in Chapter 2. For the external acoustic fluid, the equations have been introduced in Chapter 3; those for the structure in Chapter 5; and the equations for the internal acoustic fluid in Chapter 4. We refer the reader to these chapters concerning the notations used. The boundary value problem is expressed in terms of external acoustic pressure field p_E, structural displacement field $\mathbf{u}$, and internal acoustic pressure field p. For all real ω and for given $\mathbf{G}(\omega)$, $\mathbf{g}(\omega)$, $Q_E(\omega)$, and $Q(\omega)$, the problem consists in finding $p_E(\omega)$, $\mathbf{u}(\omega)$, and $p(\omega)$, such that

$$-\frac{\omega^2}{c_E^2}\,p_E - \nabla^2 p_E = i\omega\, Q_E \quad \text{in} \quad \Omega_E, \tag{6.1}$$

$$\frac{\partial p_E}{\partial \mathbf{n}^S} = \omega^2 \rho_E\, \mathbf{u}\cdot\mathbf{n}^S \quad \text{on} \quad \Gamma_E, \tag{6.2}$$

$$|p_E| = O(\frac{1}{R})\,, \quad \left|\frac{\partial p_E}{\partial R} + i\,\frac{\omega}{c_E}\,p_E\right| = O(\frac{1}{R^2})\,, \quad R = \|\mathbf{x}\| \to +\infty\,, \tag{6.3}$$

$$-\omega^2 \rho_S\, \mathbf{u} - \operatorname{div}\sigma(\mathbf{u}) = \mathbf{g} \quad \text{in} \quad \Omega_S, \tag{6.4}$$

$$\sigma(\mathbf{u})\,\mathbf{n}^S = \mathbf{G} - p_E|_{\Gamma_E}\,\mathbf{n}^S \quad \text{on} \quad \Gamma_E, \tag{6.5}$$

$$\sigma(\mathbf{u})\,\mathbf{n}^S = \mathbf{G} + p\,\mathbf{n} \quad \text{on} \quad \Gamma \cup \Gamma_Z, \tag{6.6}$$

$$-\frac{\omega^2}{\rho_0 c_0^2}\,p - i\omega\frac{\tau}{\rho_0}\nabla^2 p - \frac{1}{\rho_0}\nabla^2 p = \frac{1}{\rho_0}(i\omega Q - \tau c_0^2 \nabla^2 Q) \quad \text{in} \quad \Omega, \tag{6.7}$$

$$(1 + i\omega\,\tau)\,\frac{\partial p}{\partial \mathbf{n}} = \omega^2\,\rho_0\,\mathbf{u}\cdot\mathbf{n} + \tau\,c_0^2\,\frac{\partial Q}{\partial \mathbf{n}} \quad \text{on} \quad \Gamma, \tag{6.8}$$

$$(1 + i\omega\,\tau)\,\frac{\partial p}{\partial \mathbf{n}} = \omega^2\,\rho_0\,\mathbf{u}\cdot\mathbf{n} - i\omega\rho_0\frac{p}{Z} + \tau\,c_0^2\,\frac{\partial Q}{\partial \mathbf{n}} \quad \text{on} \quad \Gamma_Z. \tag{6.9}$$

- Eq. (6.1) is the Helmholtz equation in the external acoustic fluid (see Eq. (3.1) in which the wave number k is ω/c_E).
- Eq. (6.2) is the fluid-structure coupling condition on the external fluid-structure interface Γ_E and corresponds to the continuity of the normal velocity field of the external acoustic fluid with the normal velocity field of the structure, expressed in terms of the external acoustic pressure field p_E and the structural displacement field $\mathbf{u}$.
- Eq. (6.3) corresponds to the outward Sommerfeld radiation conditions, which means that the outward traveling waves vanish at infinity, are not reflected, and induce a loss of energy (damping induced by radiation at infinity although the external acoustic fluid be inviscid).
- Eq. (6.4) corresponds to the dynamic equation for the structure (see Eqs. (5.1) and (5.13)), in which $\{\text{div}\,\sigma(\mathbf{u})\}_i = \sigma_{ij,j}(\mathbf{u})$. The tensor $\sigma_{ij}(\mathbf{u})$ corresponds to the constitutive equation defined by Eq. (5.13),

$$\sigma_{ij}(\mathbf{u}) = a_{ijkh}(\omega)\,\varepsilon_{kh}(\mathbf{u})) + i\omega\,b_{ijkh}(\omega)\,\varepsilon_{kh}(\mathbf{u}), \tag{6.10}$$

in which tensors $a_{ijkh}(\omega)$ and $b_{ijkh}(\omega)$ are constructed following Section 5.2 and where $\varepsilon_{kh}(\mathbf{u})$ is the strain tensor defined (see

Eq. (5.3)) by

$$\varepsilon_{kh}(\mathbf{u}) = \frac{1}{2}(u_{k,h} + u_{h,k}). \qquad (6.11)$$

- Eq. (6.5) is the fluid-structure coupling condition on the external fluid-structure interface Γ_E (see Eq. (5.4)) and corresponds to the equilibrium of the surface forces on Γ_E, expressed in terms of given surface forces $\mathbf{G}$ and of the values $p_E|_{\Gamma_E}$ of the external acoustic pressure field p_E on Γ_E.

- Eq. (6.6) is the fluid-structure coupling condition on the internal fluid-structure interface $\Gamma \cup \Gamma_Z$ (see Eq. (5.6)) and corresponds to the equilibrium of the surface forces on $\Gamma \cup \Gamma_Z$, expressed in terms of given surface forces $\mathbf{G}$ and of the values p of the internal acoustic pressure field p on $\Gamma \cup \Gamma_Z$.

- Eq. (6.7) is the internal dissipative acoustic fluid equation (see Eq. (4.4)) and corresponds to the Helmholtz equation with a dissipative term and with a right-hand side member induced by the given internal acoustic source Q.

- Eq. (6.8) is the fluid-structure coupling condition on the internal fluid-structure interface Γ (see Eq. (4.5)) and corresponds to the continuity of the normal velocity field of the internal dissipative acoustic fluid with the normal velocity field of the structure and the internal acoustic source density Q, expressed in terms of the internal acoustic pressure field p and the structural displacement field $\mathbf{u}$.

- Finally, Eq. (6.9) is the fluid-structure coupling condition on the internal fluid-structure interface Γ_Z (see Eq. (4.7)).

Existence and uniqueness of the solution. Under the appropriate mathematical hypotheses, Dautray and Lions (1992), for all real ω, the boundary value problem defined by Eqs. (6.1) to (6.9), with (6.10) and (6.11), admits a unique solution in p_E, $\mathbf{u}$, and p.

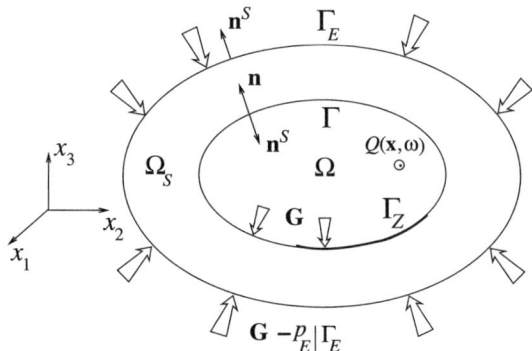

Figure 6.2 Configuration of the equivalent vibroacoustic system.

6.2 EQUIVALENT VIBROACOUSTIC BOUNDARY VALUE PROBLEM IN $(\mathbf{u}, p)$

In Eqs. (6.1) to (6.9), field p_E is eliminated in using Eq. (3.4), which is written as $p_E|_{\Gamma_E}(\omega) = p_{\text{given}}|_{\Gamma_E}(\omega) + i\omega\, \mathbf{Z}_{\Gamma_E}(\omega)\{\mathbf{u}(\omega) \cdot \mathbf{n}^S\}$. This equation allows the value $p_E|_{\Gamma_E}$ of the external acoustic pressure field p_E on the external fluid-structure interface Γ_E to be expressed as a function of the structural displacement field $\mathbf{u}$ thanks to the introduction of the acoustic impedance boundary operator $\mathbf{Z}_{\Gamma_E}(\omega)$, introduced in Chapter 3 and constructed in Chapter 10. Such elimination leads us to obtain an equivalent vibroacoustic boundary value problem in terms of $(\mathbf{u}, p)$ that constitutes the basic formulation of the vibroacoustic boundary value problem that will be used for computational issues. The configuration associated with this equivalent vibroacoustic system is shown in Figure 6.2.

The boundary value problem in terms of structural displacement field $\mathbf{u}$ and acoustic pressure field p is then written as follows. For all real ω and for given $\mathbf{G}(\omega)$, $\mathbf{g}(\omega)$, $p_{\text{given}}|_{\Gamma_E}(\omega)$, and $Q(\omega)$, find $\mathbf{u}(\omega)$ and $p(\omega)$, such that

$$-\omega^2 \rho_S\, \mathbf{u} - \operatorname{div} \sigma(\mathbf{u}) = \mathbf{g} \quad \text{in} \quad \Omega_S, \tag{6.12}$$

$$\sigma(\mathbf{u})\, \mathbf{n}^S = \mathbf{G} - p_{\text{given}}|_{\Gamma_E}\, \mathbf{n}^S - i\omega\, \mathbf{Z}_{\Gamma_E}(\omega)\{\mathbf{u} \cdot \mathbf{n}^S\}\, \mathbf{n}^S \quad \text{on} \quad \Gamma_E, \tag{6.13}$$

$$\sigma(\mathbf{u})\,\mathbf{n}^S = \mathbf{G} + p\,\mathbf{n} \quad \text{on} \quad \Gamma \cup \Gamma_Z, \tag{6.14}$$

$$-\frac{\omega^2}{\rho_0 c_0^2}\,p - i\omega\frac{\tau}{\rho_0}\nabla^2 p - \frac{1}{\rho_0}\nabla^2 p = \frac{1}{\rho_0}(i\omega Q - \tau c_0^2\nabla^2 Q) \quad \text{in} \quad \Omega, \tag{6.15}$$

$$(1 + i\omega\tau)\frac{\partial p}{\partial \mathbf{n}} = \omega^2\,\rho_0\,\mathbf{u}\cdot\mathbf{n} + \tau\,c_0^2\frac{\partial Q}{\partial \mathbf{n}} \quad \text{on} \quad \Gamma, \tag{6.16}$$

$$(1 + i\omega\tau)\frac{\partial p}{\partial \mathbf{n}} = \omega^2\,\rho_0\,\mathbf{u}\cdot\mathbf{n} - i\omega\rho_0\frac{p}{Z} + \tau\,c_0^2\frac{\partial Q}{\partial \mathbf{n}} \quad \text{on} \quad \Gamma_Z. \tag{6.17}$$

Existence and uniqueness of the solution. For all real ω, the boundary value problem defined by Eqs. (6.12) to (6.17) admits a unique solution in $\mathbf{u}$ and p which coincides with the unique solution of Eqs. (6.1) to (6.9), with Eq. (6.10) and Eq. (6.11).

6.3 CASE OF AN INTERNAL LIQUID WITH A FREE SURFACE

The configuration associated with this vibroacoustic system in presence of a free surface for the internal acoustic liquid is shown in Figure 6.3. The boundary $\partial\Omega_L$ of Ω_L is written as $\partial\Omega_L = \Gamma_0 \cup \Gamma_L \cup \Gamma_Z$ (see Figure 6.3). In such a case (see Section 4.3), the following boundary condition must be added:

$$p = 0 \quad \text{on} \quad \Gamma_0. \tag{6.18}$$

For sake of brevity, it is assumed that $\mathbf{G}$ is only applied to the liquid-structure interface $\Gamma_L \cup \Gamma_Z$. The equivalent boundary value problem in terms of structural displacement field $\mathbf{u}$ and acoustic pressure field p is then written as follows (as done in Section 6.2). For all real ω and for given $\mathbf{G}(\omega)$, $\mathbf{g}(\omega)$, $p_{\text{given}}|_{\Gamma_E}(\omega)$, and $Q(\omega)$, find $\mathbf{u}(\omega)$ and $p(\omega)$, such that

$$-\omega^2\,\rho_S\,\mathbf{u} - \operatorname{div}\sigma(\mathbf{u}) = \mathbf{g} \quad \text{in} \quad \Omega_S, \tag{6.19}$$

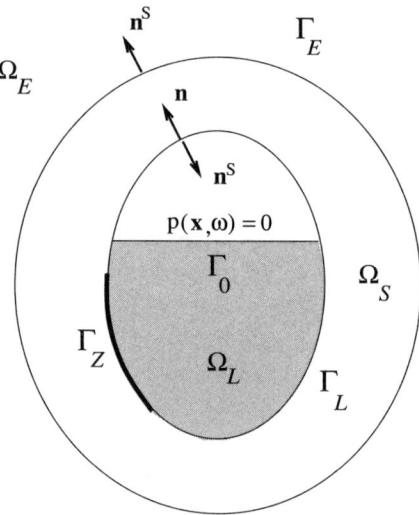

Figure 6.3 Configuration of the vibroacoustic system in the presence of a free surface for the internal acoustic liquid.

$$\sigma(\mathbf{u})\,\mathbf{n}^S = \mathbf{G} - p_{\text{given}}|_{\Gamma_E}\,\mathbf{n}^S - i\omega\,\mathbf{Z}_{\Gamma_E}(\omega)\{\mathbf{u}\cdot\mathbf{n}^S\}\,\mathbf{n}^S \quad \text{on} \quad \Gamma_E, \quad (6.20)$$

$$\sigma(\mathbf{u})\,\mathbf{n}^S = \mathbf{G} + p\,\mathbf{n} \quad \text{on} \quad \Gamma_L \cup \Gamma_Z, \quad (6.21)$$

$$-\frac{\omega^2}{\rho_0 c_0^2}\,p - i\omega\frac{\tau}{\rho_0}\nabla^2 p - \frac{1}{\rho_0}\nabla^2 p = \frac{1}{\rho_0}(i\omega Q - \tau c_0^2 \nabla^2 Q) \quad \text{in} \quad \Omega_L, \quad (6.22)$$

$$(1 + i\omega\,\tau)\,\frac{\partial p}{\partial \mathbf{n}} = \omega^2\,\rho_0\,\mathbf{u}\cdot\mathbf{n} + \tau\,c_0^2\,\frac{\partial Q}{\partial \mathbf{n}} \quad \text{on} \quad \Gamma_L, \quad (6.23)$$

$$(1 + i\omega\,\tau)\,\frac{\partial p}{\partial \mathbf{n}} = \omega^2\,\rho_0\,\mathbf{u}\cdot\mathbf{n} - i\omega\rho_0\frac{p}{Z} + \tau\,c_0^2\,\frac{\partial Q}{\partial \mathbf{n}} \quad \text{on} \quad \Gamma_Z, \quad (6.24)$$

$$p = 0 \quad \text{on} \quad \Gamma_0. \quad (6.25)$$

Existence and uniqueness of the solution. For all real ω, the boundary value problem defined by Eqs. (6.19) to (6.25) admits a unique solution in $\mathbf{u}$ and p.

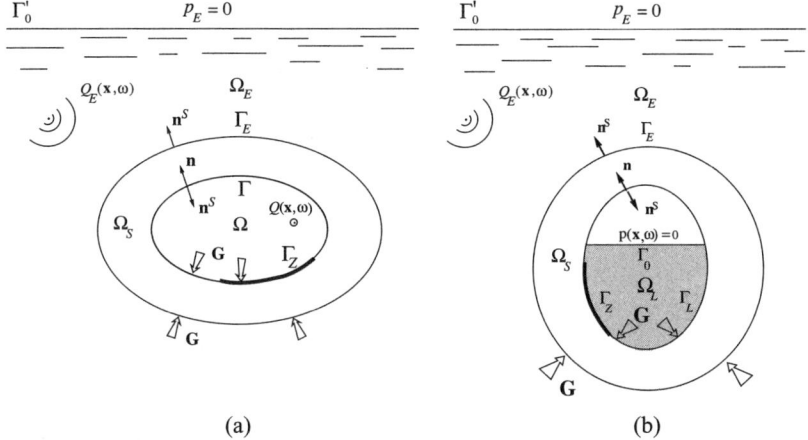

(a)   (b)

Figure 6.4 Configuration of the vibroacoustic system in presence of a free surface for the external acoustic liquid. (a) Case of an internal acoustic fluid without free surface. (b) Case of an internal acoustic liquid with free surface.

6.4 CASE OF AN EXTERNAL LIQUID WITH A FREE SURFACE

Let us consider an external acoustic liquid with a free surface Γ_0' (see Figure 6.4) on which $p_E = 0$. This problem can be transformed into a boundary value problem posed in the entire three-dimensional space (the free surface is removed) by symmetrizing the vibroacoustic system (geometry, mechanical and physical properties, boundary conditions) with respect to surface Γ_0'. The symmetrized part of the system is submitted to antisymmetric loading defined by $-\mathbf{G}(\omega)$, $-\mathbf{g}(\omega)$, $-Q_E(\omega)$, and $-Q(\omega)$. Consequently, the construction of $p_{\text{given}}|_{\Gamma_E}$ and $\mathbf{Z}_{\Gamma_E}(\omega)$, presented in Chapter 10, can be used directly to the symmetrized system defined in the entire three-dimensional space.

7

COMPUTATIONAL VIBROACOUSTIC
MODEL

The computational vibroacoustic model is constructed using the finite element discretization of the boundary value problems defined in Chapter 6. We consider a finite element mesh of structure Ω_S and a finite element mesh of internal acoustic fluid Ω. It is assumed that the two finite element meshes are compatible on interface $\Gamma \cup \Gamma_Z$. The finite element mesh of surface Γ_E is the trace of the mesh of Ω_S (see Figure 7.1). The finite element method is classically used to construct the discretization of the boundary value problem defined by Eqs. (6.12) to (6.17) or by Eqs. (6.19) to (6.25) for the case of a free surface for an internal liquid. For details concerning the practical construction of the finite element matrices of these vibroacoustic boundary value problems, the reader is referred to Ohayon and Soize (1998) (including the references). Let $\mathbb{U}(\omega)$ be the complex vector of the n_S degrees-of-freedom (DOFs) for the structure, which are the values of $\mathbf{u}(\omega)$ at the nodes of the finite element mesh of domain Ω_S. For the internal acoustic fluid, let $\mathbb{P}(\omega)$ be the complex vector of the n DOFs that are the values of $p(\omega)$ at the nodes of the finite element mesh of domain Ω. The finite element method yields the following complex matrix equations:

$$([\mathbb{A}^S(\omega)] - \omega^2[\mathbb{A}_{\mathrm{BEM}}(\omega/c_{\mathrm{E}})]) \, \mathbb{U}(\omega) + [\mathbb{C}]\,\mathbb{P}(\omega) = \mathbb{F}^S(\omega) , \quad (7.1)$$

$$([\mathbb{A}(\omega)] + [\mathbb{A}^Z(\omega)]) \, \mathbb{P}(\omega) + \omega^2\,[\mathbb{C}]^T\,\mathbb{U}(\omega) = \mathbb{F}(\omega) , \qquad (7.2)$$

50

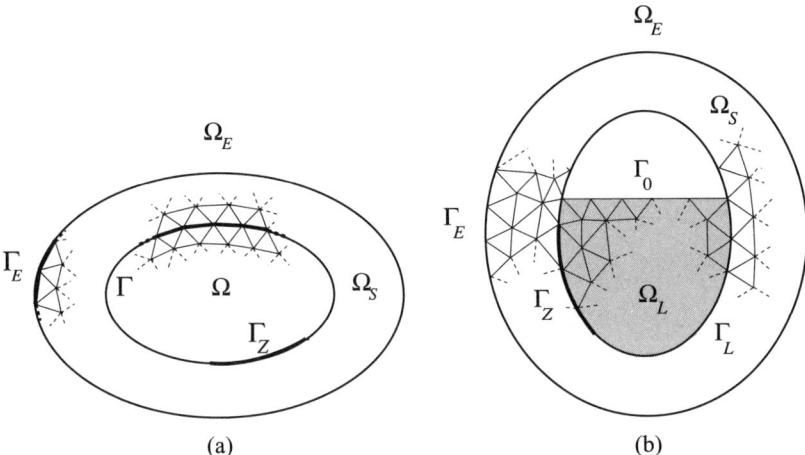

Figure 7.1 (a) Example of structure and internal fluid finite element meshes. (b) Case of an internal liquid with a free surface.

in which the matrix $[\mathbb{C}]^T$ denotes the transpose of $[\mathbb{C}]$. The matrix equation defined by Eq. (7.1) corresponds to the finite element discretization of Eq. (6.12) with the Neumann boundary conditions defined by Eqs. (6.13) and (6.14). For the case of an internal liquid with a free surface, it corresponds to Eq. (6.19) with the Neumann boundary conditions defined by Eqs. (6.20) and (6.21). The matrix equation defined by Eq. (7.2) corresponds to the finite element discretization of Eq. (6.15) with the Neumann boundary conditions defined by Eqs. (6.16) and (6.17), while, for the case of an internal liquid with a free surface, it corresponds to Eq. (6.22) with the Neumann boundary conditions defined by Eqs. (6.23) and (6.24), for which the Dirichlet condition defined by Eq. (6.25) must be taken into account setting $\mathbb{P}(\omega) = 0$ for the DOFs on Γ_0 (removing the corresponding lines and columns in the matrices). Equations (7.1) and (7.2) can be rewritten in the matrix form as

$$[\mathbb{A}_{\mathrm{FSI}}(\omega)] \begin{bmatrix} \mathbb{U}(\omega) \\ \mathbb{P}(\omega) \end{bmatrix} = \begin{bmatrix} \mathbb{F}^S(\omega) \\ \mathbb{F}(\omega) \end{bmatrix} , \qquad (7.3)$$

in which the complex matrix $[\mathbb{A}_{\text{FSI}}(\omega)]$ is defined by

$$[\mathbb{A}_{\text{FSI}}(\omega)] = \begin{bmatrix} [\mathbb{A}^S(\omega)] - \omega^2[\mathbb{A}_{\text{BEM}}(\omega/c_E)] & [\mathbb{C}] \\ \omega^2 [\mathbb{C}]^T & [\mathbb{A}(\omega)] + [\mathbb{A}^Z(\omega)] \end{bmatrix}. \quad (7.4)$$

In Eq. (7.1), the symmetric $(n_S \times n_S)$ complex matrix $[\mathbb{A}^S(\omega)]$ is defined by

$$[\mathbb{A}^S(\omega)] = -\omega^2[\mathbb{M}^S] + i\omega [\mathbb{D}^S(\omega)] + [\mathbb{K}^S(\omega)], \quad (7.5)$$

in which $[\mathbb{M}^S]$, $[\mathbb{D}^S(\omega)]$, and $[\mathbb{K}^S(\omega)]$ are symmetric $(n_S \times n_S)$ real matrices that represent the mass matrix, the damping matrix, and the stiffness matrix of the structure. Matrix $[\mathbb{M}^S]$ is positive and invertible (positive definite) and matrices $[\mathbb{D}^S(\omega)]$ and $[\mathbb{K}^S(\omega)]$ are positive and not invertible (positive semidefinite) due to the presence of six rigid body motions, since the structure has been considered as a free-free structure. In Eq. (7.2), the symmetric $(n \times n)$ complex matrix $[\mathbb{A}(\omega)]$ is defined by

$$[\mathbb{A}(\omega)] = -\omega^2[\mathbb{M}] + i\omega [\mathbb{D}] + [\mathbb{K}], \quad (7.6)$$

in which $[\mathbb{M}]$, $[\mathbb{D}]$, and $[\mathbb{K}]$ are symmetric $(n \times n)$ real matrices. From Eq. (6.7), it can easily be deduced that $[\mathbb{D}] = \tau [\mathbb{K}]$ in which τ is defined by Eq. (4.3). Matrix $[\mathbb{M}]$ is positive and invertible (positive definite). For the boundary value problem defined by Eqs. (6.12) to (6.17), matrices $[\mathbb{D}]$ and $[\mathbb{K}]$ are positive and not invertible with rank $n - 1$, while for the case of an internal liquid with a free surface for which the boundary value problem is defined by Eqs. (6.19) to (6.25), matrices $[\mathbb{D}]$ and $[\mathbb{K}]$ are positive and invertible (positive definite) due to the presence of the Dirichlet condition $\mathbb{P}(\omega) = 0$ on the free surface Γ_0. The internal fluid-structure coupling matrix $[\mathbb{C}]$ related to the coupling between the structure and the internal fluid, is a $(n_S \times n)$ real matrix that is only related to the values of $\mathbb{U}$ and $\mathbb{P}$ on the internal fluid-structure interface. The wall acoustic impedance matrix $[\mathbb{A}^Z(\omega)]$ is a symmetric $(n \times n)$ complex matrix depending on the wall acoustic impedance $Z(\mathbf{x}, \omega)$ on Γ_Z and that is only related to the values of $\mathbb{P}$ on boundary Γ_Z. The boundary element matrix $[\mathbb{A}_{\mathrm{BEM}}(\omega/c_{\mathrm{E}})]$, which depends on ω/c_{E}, is a symmetric $(n_S \times n_S)$ complex matrix that is only related to the values of $\mathbb{U}$ on the external fluid-structure interface Γ_E. This matrix is written as

$$[\mathbb{A}_{\mathrm{BEM}}(\omega/c_{\mathrm{E}})] = -\rho_{\mathrm{E}} [\mathbb{N}]^T [B_{\Gamma_E}(\omega/c_{\mathrm{E}})] [\mathbb{N}], \qquad (7.7)$$

in which $[B_{\Gamma_E}(\omega/c_{\mathrm{E}})]$ is the full symmetric $(n_E \times n_E)$ complex matrix defined in Section 10.6 and where $[\mathbb{N}]$ is a sparse $(n_E \times n_S)$ real matrix related to the finite element discretization. Complex matrix $[B_{\Gamma_E}(\omega/c_{\mathrm{E}})]$ is related to complex matrix $[Z_{\Gamma_E}(\omega)]$ by Eq. (10.31).

It should be noted that Eq. (7.3) can be solved directly ω by ω, but as explained in Chapter 1, it is recommended to use the reduced-order computational model presented in Chapter 8.

8

REDUCED-ORDER
COMPUTATIONAL MODEL

As explained at the end of Chapter 7, the computational vibroacoustic equation could be solved directly ω by ω, but, as explained in Chapter 1, it is recommended to use a reduced-order computational model, which is constructed as follows. The strategy used for constructing the reduced-order computational model consists in using the projection basis constituted of:

- The acoustic modes of the acoustic cavity with fixed boundary and without wall acoustic impedance. Two cases are considered. The first one is a closed acoustic cavity (the internal pressure varies with variation of the volume of the cavity), for which the boundary value problem has been defined in Section 6.2. For an almost closed cavity with a nonsealed wall, which are often encountered (the internal pressure does not vary with a variation of the volume of the cavity), an adapted procedure, derived from the closed cavity case, will be presented. The second one concerns an internal cavity filled with an acoustic liquid with a free surface for which the boundary value problem has been defined in Section 6.3.
- The elastic structural modes of the structure, taking into account a quasi-static effect of the internal acoustic fluid on the structure. The constitutive equation of the structure corresponds to an elastic material (see Eq. (5.35)) and consequently, the stiffness matrix of the structure has to be taken for $\omega = 0$. The static effect of the

internal acoustic fluid on the elastic structural modes in vacuo will induce an added mass matrix acting on the DOFs associated with the normal displacement of the internal fluid-structure interface. This added mass must be introduced to accelerate the convergence of the reduced-order model with respect to the acoustic modes truncation (obviously, if all the acoustic modes were taken into account, such an added mass would not be required). Nevertheless, it should be noted that in some cases, the use of modified elastic structural modes of the structure in vacuo may not be necessary and the elastic modes of the structure in vacuo would be sufficient. However, it is difficult to decide, before any computation, if the elastic modes of the structure in vacuo are enough to obtain a fast convergence with respect to the truncation of the acoustic modes. That is why it is strongly recommended to introduce such an added mass for the computation of the structural projection basis. In particular, it is absolutely required for the case of a liquid filling a closed (sealed wall) acoustic cavity.

8.1 VECTOR BASIS FOR THE INTERNAL ACOUSTIC FLUID

This step concerns the finite element calculation of a basis for the internal acoustic fluid. For the two considered cases (closed acoustic cavity including the almost closed [nonsealed wall] acoustic cavity and internal liquid with a free surface), setting $\lambda = \omega^2$, we then have to solve the following $(n \times n)$ generalized symmetric real eigenvalue problem (Bathe and Wilson, 1976),

$$[\mathbb{K}]\, \mathbb{P} = \lambda\, [\mathbb{M}]\, \mathbb{P}\,, \qquad (8.1)$$

in which the frequency-independent matrices $[\mathbb{K}]$ and $[\mathbb{M}]$ are defined in Eq. (7.6) and where $\mathbb{P}$ is the real vector of the n DOFs, which are the values of the pressure field at the nodes of the finite element mesh of the acoustic cavity.

8.1.1 Closed Acoustic Cavity

As explained in Chapter 7, matrix $[\mathbb{K}]$ is positive and noninvertible with rank $n - 1$ and consequently, there is a zero eigenvalue with multiplicity 1, denoted as λ_1 (corresponding to constant eigenvector denoted as $\mathbb{P}_1$). There is an increasing sequence of $n - 1$ strictly positive eigenvalues. Each positive eigenvalue can be multiple (case of an acoustic cavity with symmetries) and we have

$$0 = \lambda_1 < \lambda_2 \leq \ldots \leq \lambda_\alpha \leq \ldots \leq \lambda_n. \qquad (8.2)$$

Let $\mathbb{P}_1, \ldots, \mathbb{P}_\alpha, \ldots, \mathbb{P}_n$ be the eigenvectors associated with $\lambda_1, \ldots,$ $\lambda_\alpha, \ldots, \lambda_n$.

Let $1 \leq N \leq n$. We introduce the $(n \times N)$ real matrix of the constant eigenvector $\mathbb{P}_1$ and of the $N - 1$ acoustic modes $\mathbb{P}_\alpha$ associated with the first $N - 1$ strictly positive eigenvalues,

$$[\mathcal{P}] = [\mathbb{P}_1, \mathbb{P}_2 \ldots \mathbb{P}_\alpha \ldots \mathbb{P}_N]. \qquad (8.3)$$

Case of an almost closed (nonsealed wall) acoustic cavity. As explained at the beginning of this chapter, the boundary value problem defined in Section 6.2 should be modified slightly to express an acoustic fluid leak on the nonsealed wall. Such a modification would be the equivalent of writing that there is zero pressure on the nonsealed surface, which would imply that matrix $[\mathbb{K}]$ would be positive and invertible (positive definite). However, in practice, for a nonsealed wall, the part of the interface on which a zero pressure should be written is unknown. Consequently, the following modeling can be done. The formulation given in Section 8.1.1 (for a closed acoustic cavity) is kept and the nonsealed wall effects are taken into account by retaining only the eigenvectors associated with the positive eigenvalues computed in Section 8.1.1,

$$\lambda_2 \leq \ldots \leq \lambda_\alpha \leq \ldots \leq \lambda_n. \qquad (8.4)$$

Therefore, the eigenvectors $\mathbb{P}_2, \ldots, \mathbb{P}_\alpha, \ldots, \mathbb{P}_n$ with $\lambda_2, \ldots, \lambda_\alpha, \ldots, \lambda_n$ constitute a basis. Let $1 \leq N \leq n - 1$. We introduce the $(n \times N)$ real matrix of the N acoustic modes $\mathbb{P}_\alpha$ associated with the first N strictly positive eigenvalues,

$$[\mathcal{P}] = [\mathbb{P}_2 \dots \mathbb{P}_\alpha \dots \mathbb{P}_{N+1}]. \tag{8.5}$$

8.1.2 Internal Liquid with a Free Surface

As explained in Chapter 7, matrix $[\mathbb{K}]$ is positive and invertible (positive definite) due to the presence of zero pressure on the free surface and consequently, there is an increasing sequence of n strictly positive eigenvalues. Each positive eigenvalue can be multiple (case of an acoustic cavity with symmetries) and we have

$$\lambda_1 \le \dots \le \lambda_\alpha \le \dots \le \lambda_n. \tag{8.6}$$

Let $\mathbb{P}_1, \dots, \mathbb{P}_\alpha, \dots, \mathbb{P}_n$ be the eigenvectors associated with $\lambda_1, \dots,$ $\lambda_\alpha, \dots, \lambda_n$ and called the acoustic modes. Let $1 \le N \le n$. We introduce the $(n \times N)$ real matrix of the N acoustic modes $\mathbb{P}_\alpha$ associated with the first N strictly positive eigenvalues,

$$[\mathcal{P}] = [\mathbb{P}_1 \dots \mathbb{P}_\alpha \dots \mathbb{P}_N]. \tag{8.7}$$

8.1.3 Orthogonality Properties

The following classical orthogonality properties hold:

$$[\mathcal{P}]^T [\mathbb{M}] [\mathcal{P}] = [M], \tag{8.8}$$

$$[\mathcal{P}]^T [\mathbb{K}] [\mathcal{P}] = [K], \tag{8.9}$$

in which $[\mathcal{P}]^T$ is the transpose of $[\mathcal{P}]$, where $[M]$ is a diagonal matrix of positive real numbers such that $[M]_{\alpha\beta} = \mu_\alpha \, \delta_{\alpha\beta}$, and where $[K]$ is the diagonal matrix of the eigenvalues such that $[K]_{\alpha\beta} = \mu_\alpha \lambda_\alpha \, \delta_{\alpha\beta}$ (for the nonzero eigenvalues, the eigenfrequencies are $\omega_\alpha = \sqrt{\lambda_\alpha}$). Consequently, $[M]$ is a positive-definite matrix and matrix $[K]$ is positive definite for an almost closed (nonsealed wall) acoustic cavity and for an internal liquid with a free surface on which the pressure is zero, while $[K]$ is only positive (noninvertible) for a closed acoustic cavity. For the case of an almost closed (nonsealed wall) acoustic cavity, the equation $[K]_{\alpha\beta} = \mu_\alpha \, \lambda_\alpha \, \delta_{\alpha\beta}$ must be replaced by $[K]_{\alpha\beta} = \mu_\alpha \, \lambda_{\alpha+1} \, \delta_{\alpha\beta}$.

8.2 VECTOR BASIS FOR THE ELASTIC STRUCTURE

As explained in the beginning of this chapter, an appropriate basis for the elastic structure is constituted of the elastic structural modes of the structure taking into account a quasi-static effect of the internal acoustic fluid on the structure. In the first step, we will present the computation of the usual elastic structural modes of the structure in vacuo; see Bathe and Wilson (1976); Geradin and Rixen (1997). The second step will be devoted to the computation of the appropriate basis for the elastic structure.

8.2.1 Computation of the Elastic Structural Modes of the Structure in Vacuo

This step concerns the finite element calculation of the elastic structural modes for structure Ω_S in vacuo, for which the constitutive equation corresponds to an elastic material. Setting $\lambda^S = \omega^2$, we then have the following $(n_S \times n_S)$ generalized symmetric real eigenvalue problem:

$$[\mathbb{K}^S(0)] \, \mathbb{U} = \lambda^S [\mathbb{M}^S] \, \mathbb{U}. \qquad (8.10)$$

It can be shown that there is a zero eigenvalue with multiplicity 6 (corresponding to the six rigid body motions) and that there is an increasing sequence of $n_S - 6$ strictly positive eigenvalues; each positive eigenvalue can be multiple (case of a structure with symmetries),

$$0 < \lambda_1^S \le \ldots \le \lambda_\alpha^S \le \ldots . \qquad (8.11)$$

Let $\mathbb{U}_1, \ldots, \mathbb{U}_\alpha, \ldots$ be the eigenvectors (called the elastic structural modes) associated with $\lambda_1^S, \ldots, \lambda_\alpha^S, \ldots$. Let $0 < N_S \leq n_S - 6$. We introduce the $(n_S \times N_S)$ real matrix of the N_S elastic structural modes $\mathbb{U}_\alpha$ associated with the first N_S strictly positive eigenvalues,

$$[\mathcal{U}] = [\,\mathbb{U}_1 \ldots \mathbb{U}_\alpha \ldots \mathbb{U}_{N_S}\,]. \tag{8.12}$$

One has the orthogonality properties,

$$[\mathcal{U}]^T [\mathbb{M}^S][\mathcal{U}] = [M^S], \tag{8.13}$$

$$[\mathcal{U}]^T [\mathbb{K}^S(0)][\mathcal{U}] = [K^S(0)], \tag{8.14}$$

in which $[M^S]$ is a diagonal matrix of positive real numbers such that $[M^S]_{\alpha\beta} = \mu_\alpha^S \delta_{\alpha\beta}$ and where $[K^S(0)]$ is the diagonal matrix of the eigenvalues such that $[K^S(0)]_{\alpha\beta} = \mu_\alpha^S \lambda_\alpha^S \delta_{\alpha\beta}$ (the eigenfrequencies are $\omega_\alpha^S = \sqrt{\lambda_\alpha^S}$).

8.2.2 Computation of an Appropriate Basis for the Elastic Structure

For the computation of the appropriate basis for the elastic structure, the practical numerical procedure is given in the following. The justification of this numerical procedure is presented in Section 8.4 (local equations and corresponding finite element discretization). Equation (8.10) is then replaced by the following one,

$$[\mathbb{K}^S(0)]\,\mathbb{U} = \lambda^S\,([\mathbb{M}^S] + [\mathbb{M}^A])\,\mathbb{U}, \tag{8.15}$$

in which $[\mathbb{M}^A]$ is the positive symmetric $(n_S \times n_S)$ real matrix, called the added mass matrix, which corresponds to the quasi-static effect of the internal acoustic fluid on the structure and for which the construction is explained below. It should be noted that the nonzero elements of matrix $[\mathbb{M}^A]$ are only related to the DOFs of the internal fluid-structure interface. As in Section 8.2.1, there is a zero eigenvalue with multiplicity 6 (corresponding to the six rigid body motions) and an increasing sequence of $n_S - 6$ strictly positive eigenvalues, each positive eigenvalue being able to be multiple (case of a structure with symmetries),

$$0 < \lambda_1^S \leq \ldots \leq \lambda_\alpha^S \leq \ldots . \tag{8.16}$$

The eigenvectors $\mathbb{U}_1, \ldots, \mathbb{U}_\alpha, \ldots$ associated with $\lambda_1^S, \ldots, \lambda_\alpha^S, \ldots$ constitute a basis for the elastic structure. As previously, for $0 < N_S \leq n_S - 6$, we introduce the $(n_S \times N_S)$ real matrix of the N_S basis vectors $\mathbb{U}_\alpha$ associated with the first N_S strictly positive eigenvalues,

$$[\mathcal{U}] = [\mathbb{U}_1 \ldots \mathbb{U}_\alpha \ldots \mathbb{U}_{N_S}] . \tag{8.17}$$

The following orthogonality properties hold:

$$[\mathcal{U}]^T ([\mathbb{M}^S] + [\mathbb{M}^A]) [\mathcal{U}] = [M^{SA}] , \tag{8.18}$$

$$[\mathcal{U}]^T [\mathbb{K}^S(0)] [\mathcal{U}] = [K^S(0)] , \tag{8.19}$$

in which $[M^{SA}]$ is a diagonal matrix of positive real numbers such that $[M^{SA}]_{\alpha\beta} = \mu_\alpha^S \delta_{\alpha\beta}$ and where $[K^S(0)]$ is the diagonal matrix of the eigenvalues such that $[K^S(0)]_{\alpha\beta} = \mu_\alpha^S \lambda_\alpha^S \delta_{\alpha\beta}$. It should be noted that $\omega_\alpha^S = \sqrt{\lambda_\alpha^S}$ is not an eigenfrequency of the structure in vacuo but is an eigenfrequency of the structure with an added mass effect induced by the internal acoustic fluid. For a closed acoustic cavity and for an internal liquid with a free surface, the construction of matrix $[\mathbb{M}^A]$ is given in the following.

Construction of added mass matrix $[\mathbb{M}^A]$ for a closed acoustic cavity. For a closed acoustic cavity, the introduced added mass does not correspond to an incompressible fluid and consequently, the deformations of the interface inducing a volume variation of the cavity are authorized. The positive symmetric $(n_S \times n_S)$ real matrix $[\mathbb{M}^A]$ is written as

$$[\mathbb{M}^A] = [\mathbb{C}] [\mathbb{S}] [\mathbb{C}]^T , \tag{8.20}$$

in which $[\mathbb{C}]$ is the $(n_S \times n)$ real matrix defined in Eq. (7.4) and where $[\mathbb{S}]$ is the positive symmetric $(n \times n)$ real matrix, which is constructed in solving the following linear matrix equation,

$$[\mathbb{K}][\mathbb{S}] = [I_n], \tag{8.21}$$

under the constraint matrix equation

$$[\mathbb{B}]^T [\mathbb{S}] = [0], \tag{8.22}$$

in which matrix $[\mathbb{K}]$ is defined in Eq. (7.6), $[I_n]$ is the $(n \times n)$ identity matrix, $[0]$ is the $(1 \times n)$ zero matrix, and where $[\mathbb{B}]$ is the $(n \times 1)$ real matrix constructed by

$$[\mathbb{B}] = \rho_0 c_0^2 [\mathbb{M}][\mathbb{1}], \tag{8.23}$$

with matrix $[\mathbb{M}]$ defined in Eq. (7.6) and where $[\mathbb{1}]$ is the $(n \times 1)$ matrix such that $[\mathbb{1}]_{j1} = 1$. It should be noted that the numerical construction of matrix $[\mathbb{M}^A]$ can be viewed as the result of a Schur complement calculation with the constraint defined by Eq. (8.22).

For the justification of this construction, the interested reader is referred to Section 8.4.

Construction of added mass matrix $[\mathbb{M}^A]$ for an internal liquid with a free surface. For an internal liquid with a free surface, the introduced added mass corresponds to an incompressible fluid. The positive symmetric $(n_S \times n_S)$ real matrix $[\mathbb{M}^A]$ is written as

$$[\mathbb{M}^A] = [\mathbb{C}][\mathbb{S}][\mathbb{C}]^T, \tag{8.24}$$

in which $[\mathbb{C}]$ is the $(n_S \times n)$ real matrix defined in Eq. (7.4) and where $[\mathbb{S}]$ is the positive symmetric $(n \times n)$ real matrix that is constructed in solving the following linear matrix equation

$$[\mathbb{K}][\mathbb{S}] = [I_n], \tag{8.25}$$

in which matrix $[\mathbb{K}]$ is defined in Eq. (7.6). It should be noted that the numerical construction of matrix $[\mathbb{M}^A]$ can be viewed as the result of a Schur complement calculation.

For the justification of this construction, the interested reader is referred to Section 8.5.

8.3 CONSTRUCTION OF THE REDUCED-ORDER COMPUTATIONAL MODEL

The reduced-order computational model, of order $N_S \ll n_S$ and $N \ll n$, is obtained by projecting Eqs. (7.1) and (7.2) as follows:

$$\mathbb{U}(\omega) = [\mathcal{U}]\,\mathbf{q}^S(\omega)\,, \qquad (8.26)$$

$$\mathbb{P}(\omega) = [\mathcal{P}]\,\mathbf{q}(\omega)\,, \qquad (8.27)$$

in which matrix $[\mathcal{U}]$ is constructed in Section 8.2.2 and $[\mathcal{P}]$ is constructed in Section 8.2.1.

Concerning the choice of the structural basis, we use the appropriate basis, which takes into account the introduced added mass effect (Section 8.2.2). In particular, for liquids, such a choice is highly recommended because the choice of the elastic modes of the structure in vacuo (Section 8.2.1) requires a very large number of acoustic modes to reach a convergence of the reduced-order computational model (see, for instance, David and Menelle, 2007), for an experimental analysis and validation). Such a situation could also occur for an acoustic cavity filled with a gas and coupled with a light structure. Nevertheless, for a standard vibroacoustic system made up of a metallic structure coupled with an internal air cavity, the basis constituted of the elastic modes of the structure in vacuo (Section 8.2.1) is often sufficient to obtain a rapid convergence. The complex vectors $\mathbf{q}^S(\omega)$ and $\mathbf{q}(\omega)$ of dimension N_S and N are the solution of the following equation:

$$[A_{\mathrm{FSI}}(\omega)] \begin{bmatrix} \mathbf{q}^S(\omega) \\ \mathbf{q}(\omega) \end{bmatrix} = \begin{bmatrix} \mathbf{f}^S(\omega) \\ \mathbf{f}(\omega) \end{bmatrix}\,, \qquad (8.28)$$

in which the complex matrix $[A_{\mathrm{FSI}}(\omega)]$ is defined by

$$\begin{bmatrix} [A^S(\omega)] - \omega^2[A_{\mathrm{BEM}}(\omega/c_{\mathrm{E}})] & [C] \\ \omega^2\,[C]^T & [A(\omega)] + [A^Z(\omega)] \end{bmatrix}. \qquad (8.29)$$

In Eq. (8.29), the symmetric $(N_S \times N_S)$ complex matrix $[A^S(\omega)]$ is defined by

$$[A^S(\omega)] = -\omega^2[M^S] + i\omega\,[D^S(\omega)] + [K^S(\omega)]\,, \qquad (8.30)$$

in which $[M^S]$, $[D^S(\omega)]$, and $[K^S(\omega)]$ are positive-definite symmetric $(N_S \times N_S)$ real matrices such that $[M^S] = [\mathcal{U}]^T\,[\mathbb{M}^S]\,[\mathcal{U}]$, $[D^S(\omega)] = [\mathcal{U}]^T\,[\mathbb{D}^S(\omega)]\,[\mathcal{U}]$, and $[K^S(\omega)] = [\mathcal{U}]^T\,[\mathbb{K}^S(\omega)]\,[\mathcal{U}]$. The symmetric $(N \times N)$ complex matrix $[A(\omega)]$ is defined by

$$[A(\omega)] = -\omega^2[M] + i\omega\,[D] + [K]\,, \qquad (8.31)$$

in which $[M]$, $[D]$, and $[K]$ are diagonal $(N \times N)$ real matrices. Matrix $[M]$ is positive and invertible. The diagonal $(N \times N)$ real matrix $[D]$ is written as $[D] = \tau\,[K]$, in which τ is defined by Eq. (4.3). The $(N_S \times N)$ real matrix $[C]$ is written as $[C] = [\mathcal{U}]^T\,[\mathbb{C}]\,[\mathcal{P}]$. The symmetric $(N \times N)$ complex matrix $[A^Z(\omega)]$ is such that $[A^Z(\omega)] = [\mathcal{P}]^T\,[\mathbb{A}^Z(\omega)]\,[\mathcal{P}]$, and finally, the symmetric $(N_S \times N_S)$ complex matrix $[A_{\mathrm{BEM}}(\omega/c_{\mathrm{E}})]$ is given by $[A_{\mathrm{BEM}}(\omega/c_{\mathrm{E}})] = [\mathcal{U}]^T\,[\mathbb{A}_{\mathrm{BEM}}(\omega/c_{\mathrm{E}})]\,[\mathcal{U}]$. The given generalized excitations are written as $\mathbf{f}^S(\omega) = [\mathcal{U}]^T\,\mathbb{F}^S(\omega)$ and $\mathbf{f}(\omega) = [\mathcal{P}]^T\,\mathbb{F}(\omega)$.

8.4 QUASI-STATIC EFFECTS FOR A CLOSED ACOUSTIC CAVITY

For constructing the added mass matrix $[\mathbb{M}^A]$ defined by Eqs. (8.20) to (8.23) for a closed acoustic cavity (which does not correspond to an incompressible fluid and, consequently, for which the deformations of the internal fluid-structure interface inducing a volume variation of the cavity are authorized), we need to solve the internal acoustic equations of Chapter 4 for $\omega = 0$, under prescribed normal wall displacement and without any dissipation, wall acoustic impedance, and acoustic sources. To obtain the equations, the acoustic velocity field $\mathbf{v}$ is expressed in terms of the acoustic displacement field $\mathbf{u}_F$ such that $\mathbf{v} = i\omega\,\mathbf{u}_F$. Substituting this last equation in Eqs. (4.1) and (4.2) and using the boundary condition defined in Section 4.2, $\mathbf{v} \cdot \mathbf{n} = i\omega\,\mathbf{u}$ on $\partial\Omega = \Gamma \cup \Gamma_Z$ in which $\mathbf{u}$ is the structural displacement, we obtain the following equations in domain Ω:

$$\nabla p - \rho_0 \omega^2 \mathbf{u}_F = \mathbf{0} \tag{8.32}$$

$$p + \rho_0 c_0^2 \nabla \cdot \mathbf{u}_F = 0, \tag{8.33}$$

with the boundary condition on $\partial\Omega$,

$$\mathbf{u}_F \cdot \mathbf{n} = \mathbf{u} \cdot \mathbf{n}. \tag{8.34}$$

Setting $\omega = 0$ in Eqs. (8.32) to (8.34) and adding the condition $\nabla \times \mathbf{u}_F = \mathbf{0}$ corresponding to an acoustic fluid that is irrotational at zero frequency, we obtain, in domain Ω,

$$\nabla p = \mathbf{0} \tag{8.35}$$

$$p + \rho_0 c_0^2 \nabla \cdot \mathbf{u}_F = 0 \tag{8.36}$$

$$\nabla \times \mathbf{u}_F = \mathbf{0}, \tag{8.37}$$

with the boundary condition on $\partial\Omega$,

$$\mathbf{u}_F \cdot \mathbf{n} = \mathbf{u} \cdot \mathbf{n}. \tag{8.38}$$

For given $\mathbf{u} \cdot \mathbf{n}$ on $\partial\Omega$, deducing from Eq. (8.35) that p is independent of $\mathbf{x}$, integrating Eq. (8.36) in Ω, using the Stokes theorem and Eq. (8.38), it can be deduced that the unique solution in p of Eqs. (8.35), (8.36), and (8.38), denoted by p_0, is written as

$$p_0 = -\frac{\rho_0 c_0^2}{|\Omega|} \int_{\partial\Omega} \mathbf{u} \cdot \mathbf{n}, \tag{8.39}$$

in which $|\Omega| = \int_{\Omega} d\mathbf{x}$ is the volume of domain Ω. Substituting Eq. (8.39) into Eq. (8.36), we obtain the equations in $\mathbf{u}_F$ that are written in domain Ω as

$$\nabla \cdot \mathbf{u}_F = \frac{1}{|\Omega|} \int_{\partial\Omega} \mathbf{u} \cdot \mathbf{n}, \tag{8.40}$$

$$\nabla \times \mathbf{u}_F = \mathbf{0}, \tag{8.41}$$

with the boundary condition on $\partial\Omega$,

$$\mathbf{u}_F \cdot \mathbf{n} = \mathbf{u} \cdot \mathbf{n}. \qquad (8.42)$$

Assuming that Ω is a simply connected domain, for satisfying Eq. (8.41), a displacement potential field $\varphi(\mathbf{x})$, defined up to an additive constant, is introduced such that $\mathbf{u}_F(\mathbf{x}) = \nabla\varphi(\mathbf{x})$. To ensure the uniqueness of φ, the following constraint is imposed,

$$\int_{\Omega} \varphi(\mathbf{x})\, d\mathbf{x} = 0. \qquad (8.43)$$

The equation in domain Ω can then be written as

$$\nabla^2\varphi = \frac{1}{|\Omega|} \int_{\partial\Omega} \mathbf{u} \cdot \mathbf{n}, \qquad (8.44)$$

with the boundary condition on $\partial\Omega$,

$$\frac{\partial\varphi}{\partial\mathbf{n}} = \mathbf{u} \cdot \mathbf{n}. \qquad (8.45)$$

Concerning the finite element approximation of Eqs. (8.43) to (8.45), let Φ be the real vector of dimension n corresponding to the finite element discretization of field φ. The finite element discretization of Eq. (8.43) yields $[\mathbb{B}]^T \Phi = \mathbf{0}$ in which matrix $[\mathbb{B}]$ is defined by Eq. (8.23). The finite element discretization of Eq. (8.44) with Eq. (8.45) yields

$$\rho_0\, [\mathbb{K}]\, \Phi = -[\mathbb{C}]^T \mathbb{U}. \qquad (8.46)$$

The matrix $[\mathbb{M}^A]$ is such that

$$-\omega^2\, [\mathbb{M}^A]\, \mathbb{U} = \omega^2\, \rho_0\, [\mathbb{C}]\, \Phi. \qquad (8.47)$$

Then, eliminating Φ between Eqs. (8.46) and (8.47) yields Eq. (8.20), in which matrix $[\mathbb{S}]$ is calculated solving the linear matrix Eq. (8.21) under the constraint defined by Eq. (8.22).

8.5 QUASI-STATIC EFFECTS FOR AN INTERNAL LIQUID WITH A FREE SURFACE

For constructing the added mass matrix $[\mathbb{M}^A]$ defined by Eqs. (8.24) and (8.25) for an internal liquid with a free surface (which corresponds to an incompressible fluid), the equations established in Section 8.4 are adapted. For that, Eq. (8.43) is removed and replaced by the free surface equation defined by Eq. (4.8) expressed in terms of φ. The following equations in domain Ω are then written as

$$\mathbf{\nabla}^2 \varphi = 0, \tag{8.48}$$

with the boundary condition on $\partial\Omega_L\backslash\Gamma_0$,

$$\frac{\partial\varphi}{\partial\mathbf{n}} = \mathbf{u}\cdot\mathbf{n}, \tag{8.49}$$

and the free surface boundary condition on Γ_0,

$$\varphi = 0. \tag{8.50}$$

The finite element discretization of Eq. (8.48) with Eqs. (8.49) and (8.50) yields

$$\rho_0\,[\mathbb{K}]\,\Phi = -[\mathbb{C}]^T\,\mathbb{U}. \tag{8.51}$$

Matrix $[\mathbb{M}^A]$ is such that

$$-\omega^2\,[\mathbb{M}^A]\,\mathbb{U} = \omega^2\,\rho_0\,[\mathbb{C}]\,\Phi. \tag{8.52}$$

Then, eliminating Φ between Eqs. (8.51) and (8.52) yields Eq. (8.24), in which matrix $[\mathbb{S}]$ is calculated solving the linear matrix Eq. (8.25).

UNCERTAINTY QUANTIFICATION IN COMPUTATIONAL VIBROACOUSTICS

9.1 UNCERTAINTY AND VARIABILITY

The *designed vibroacoustic system* is used to manufacture the *real vibroacoustic system* and to construct the nominal computational vibroacoustic model (also called the *mean computational vibroacoustic model* or sometime the *mean model*) using a mathematical-mechanical modeling process for which the main objective is the prediction of the responses of the real vibroacoustic system. This system can exhibit a variability in its responses due to fluctuations in the manufacturing process and due to small variations of the configuration around a nominal configuration associated with the designed vibroacoustic system. The mean computational model that results from a mathematical-mechanical modeling process of the designed vibroacoustic system has parameters (such as geometry, mechanical properties, and boundary conditions) that can be uncertain (for example, parameters related to the structure, the internal acoustic fluid, the wall acoustic impedance). In this case, there are *uncertainties on the computational vibroacoustic model parameters*, also called *uncertainties on the system parameters*. On the other hand, the modeling process induces some *modeling errors* defined as the *model uncertainties*. Figure 9.1 summarizes the two types of uncertainties in a computational model and the variabilities of a real system. It is important to take into account both the uncertainties on the computational vibroacoustic model parameters and the model

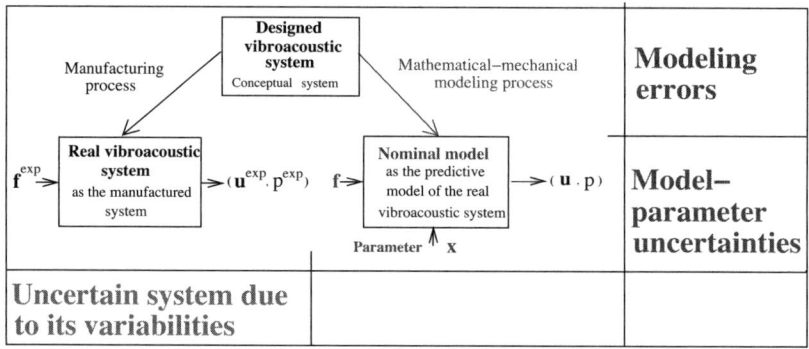

Figure 9.1 Variabilities and types of uncertainties in computational vibroacoustics.

uncertainties to improve the predictions in order to use such a computational vibroacoustic model to carry out robust optimization, robust
design, and robust updating with respect to uncertainties. Today, it is
well understood that, as soon as the probability theory can be used,
then the stochastic approach of uncertainties is the most powerful, efficient, and effective tool for modeling and for solving direct problems
and inverse problems related to the identification. The developments
presented in the following are carried out within the framework of the
probability theory.

9.2 PROBABILISTIC APPROACH OF
SYSTEM-PARAMETER UNCERTAINTIES

The *parametric probabilistic approach* consists in modeling the *uncertain parameters of the computational vibroacoustic model* by random
variables and then in constructing the stochastic model of these random variables using the available information. Such an approach is
very well adapted and very efficient to take into account the uncertainties in the computational model parameters. Many works have
been published and a state-of-the-art can be found, for instance, in
Ghanem and Spanos (1991, 2003); Mace et al. (2005); Schuëller (2005,
2007); and Deodatis and Spanos (2008).

9.3 PROBABILISTIC APPROACH OF MODELING ERRORS

Concerning *model uncertainties* induced by *modeling errors*, it is well understood that the prior and posterior probability models of the uncertain parameters of the computational model are not sufficient and do not have the capability to take into account model uncertainties in the context of computational mechanics as explained, for instance, in Beck and Katafygiotis (1998) and in Soize (2000, 2001, 2005b). Two main methods can be used to take into account model uncertainties (modeling errors).

9.3.1 Output-Prediction-Error Method

If experiments are available, the observed prediction error is the difference between the measured real system output and the computational model output. It then consists in introducing a stochastic model of a noise that is the difference between the real system output and the computational model output. A posterior probability model can then be constructed (Beck and Katafygiotis, 1998; Beck and Au, 2002) using the Bayesian method (Spall, 2003; Kaipio and Somersalo, 2005). Such an approach is efficient but requires experimental data. In this case, the posterior probability model of the uncertain parameters of the computational model strongly depends on the probability model of the noise that is added to the model output and is often unknown. In addition, for many problems, it can be necessary to take into account the modeling errors at the operators level of the mean computational model. For instance, such an approach seems to be necessary to take into account the modeling errors on the mass and the stiffness operators of a computational dynamical model in order to analyze the generalized eigenvalue problem. It is also the case for the robust design optimization performed with an uncertain computational model for which the design parameters of the computational model are not fixed but vary inside an admissible set of values.

If there are no experimental data, then this method cannot be used because there is generally no information concerning the probability model of the noise that is added to the computational model output.

9.3.2 Nonparametric Probabilistic Method

The *nonparametric probabilistic approach* of model uncertainties induced by modeling errors is an approach proposed in Soize (2000) that is an alternative method to the previous output-prediction-error method and allows modeling errors to be taken into account at the operators' level by introducing random operators and not at the model output level through the introduction of an additive noise. It should be noted that this second approach allows a prior probability model of model uncertainties to be constructed even if no experimental data are available. This nonparametric probabilistic approach is based on the use of a reduced-order model and the random matrix theory. It consists in directly constructing the stochastic modeling of the operators of the mean computational model. The random matrix theory (Mehta, 1991) and its developments in the context of dynamics, vibration, and acoustics (Soize, 2000, 2001, 2005b, 2010b; Wright and Weaver, 2010) are used to construct the prior probability distribution of the random matrices modeling the uncertain operators of the mean computational model. This prior probability distribution is constructed by using the maximum entropy principle (Jaynes, 1957), in the context of information theory (Shannon, 1948), for which the constraints are defined by the available information (Soize, 2000, 2001, 2003a, 2005a,b, 2010b). Since the basic paper Soize (2000), many works have been published in order

- To validate, using experimental results, the nonparametric probabilistic approach of both the system-parameter uncertainties and the model uncertainties induced by modeling errors

(Chebli and Soize, 2004; Soize, 2005b; Chen et al., 2006; Duchereau and Soize, 2006; Soize et al., 2008a; Durand et al., 2008; Fernandez et al., 2009, 2010);

- To extend the applicability of the theory to the case of nonhomogeneous uncertainties using substructuring techniques based on the Craig and Bampton procedure (Soize and Chebli, 2003; Chebli and Soize, 2004; Mignolet et al., 2013) and to various aplications (Soize, 2003b; Capiez-Lernout and Soize, 2004; Desceliers et al., 2004; Capiez-Lernout et al., 2005; Cottereau et al., 2007; Soize, 2008; Das and Ghanem, 2009; Kassem et al., 2009);

- To extend the theory to new ensembles of positive-definite random matrices yielding a more flexible description of the dispersion levels (Mignolet and Soize, 2008a);

- To apply the theory for analyzing complex dynamical systems in medium-frequency range, including vibroacoustic systems (Ghanem and Sarkar, 2003; Soize, 2003b; Chebli and Soize, 2004; Capiez-Lernout et al., 2006; Duchereau and Soize, 2006; Arnst et al., 2006; Durand et al., 2008; Pellissetti et al., 2008; Desceliers et al., 2009; Fernandez et al., 2009, 2010; Kassem et al., 2011; Soize, 2012a);

- To analyze nonlinear dynamical systems (i) with local nonlinear elements (Desceliers et al., 2004; Sampaio and Soize, 2007a,b; Batou and Soize, 2009a,b; Ritto et al., 2009, 2010; Wang et al., 2011) and (ii) with nonlinear geometrical effects (Mignolet and Soize, 2008b; Capiez-Lernout et al., 2012).

Concerning the coupling of the parametric probabilistic approach of uncertain computational model parameters with the nonparametric probabilistic approach of model uncertainties induced by modeling errors, a methodology has recently been proposed (Soize, 2010a; Batou et al., 2011). This *generalized probabilistic approach* of uncertainties in computational dynamics uses the random matrix theory. The proposed approach allows the prior probability model of each type of uncertainties (uncertainties on the computational model parameters and model uncertainties) to be separately constructed and identified.

Robust updating or robust design optimization consists in updating a computational model or in optimizing the design of a mechanical system with a computational model and in taking into account the uncertainties in the computational model parameters and the modeling uncertainties. An overview of the computational methods in optimization considering uncertainties can be found in Schuëller and Jensen (2008). Robust updating and robust design developments with uncertainties in the computational model parameters are developed in Papadimitriou et al. (2001); Taflanidis and Beck (2008); and Goller et al. (2009), while robust updating and robust design optimization with modeling uncertainties can be found in Capiez-Lernout and Soize (2008a,b,c); Soize et al. (2008b); and Ritto et al. (2010).

9.4 UNCERTAINTIES AND STOCHASTIC REDUCED-ORDER VIBROACOUSTIC MODEL

This section is devoted to the construction of the stochastic reduced-order vibroacoustic model of both the system-parameter uncertainties and the modeling errors using the nonparametric probabilistic approach and random matrix theory (for the details, see Durand et al., 2008; Soize, 2010b, 2012a,b). We apply this methodology to the reduced-order computational vibroacoustic model defined in Section 8.3. It is assumed that there are no uncertainties in the boundary element matrix $[A_{\text{BEM}}(\omega/c_{\text{E}})]$ and in the wall acoustic impedance matrix $[A^Z(\omega)]$. Consequently, for fixed values N_S and N, the stochastic reduced-order computational structural-acoustic model of order N_S and N is written as

$$\mathbf{U}(\omega) = [\mathcal{U}]\,\mathbf{Q}^S(\omega)\,, \qquad (9.1)$$

$$\mathbf{P}(\omega) = [\mathcal{P}]\,\mathbf{Q}(\omega)\,, \qquad (9.2)$$

in which, for all fixed ω, the complex random vectors $\mathbf{Q}^S(\omega)$ and $\mathbf{Q}(\omega)$ of dimension N_S and N are the solution of the following equation:

$$[\mathbf{A}_{\mathrm{FSI}}(\omega)] \begin{bmatrix} \mathbf{Q}^S(\omega) \\ \mathbf{Q}(\omega) \end{bmatrix} = \begin{bmatrix} \mathbf{f}^S(\omega) \\ \mathbf{f}(\omega) \end{bmatrix} , \qquad (9.3)$$

and where the complex random matrix $[\mathbf{A}_{\mathrm{FSI}}(\omega)]$ is written as

$$\begin{bmatrix} [\mathbf{A}^S(\omega)] - \omega^2[A_{\mathrm{BEM}}(\omega/c_{\mathrm{E}})] & [\mathbf{C}] \\ \omega^2 [\mathbf{C}]^T & [\mathbf{A}(\omega)] + [A^Z(\omega)] \end{bmatrix}. \qquad (9.4)$$

The symmetric $(N_S \times N_S)$ complex random matrix $[\mathbf{A}^S(\omega)]$ is defined by

$$[\mathbf{A}^S(\omega)] = -\omega^2[\mathbf{M}^S] + i\omega [\mathbf{D}^S(\omega)] + [\mathbf{K}^S(\omega)], \qquad (9.5)$$

in which the positive-definite symmetric $(N_S \times N_S)$ real matrices $[\mathbf{M}^S]$, $[\mathbf{D}^S(\omega)]$, and $[\mathbf{K}^S(\omega)]$ are random matrices whose probability distributions are constructed in Sections 9.6 and 9.7. The symmetric $(N \times N)$ complex random matrix $[\mathbf{A}(\omega)]$ is written as

$$[\mathbf{A}(\omega)] = -\omega^2[\mathbf{M}] + i\omega [\mathbf{D}] + [\mathbf{K}], \qquad (9.6)$$

in which $[\mathbf{M}]$, $[\mathbf{D}]$, and $[\mathbf{K}]$ are symmetric $(N \times N)$ real random matrices. Random matrix $[\mathbf{M}]$ is positive definite. For a closed (sealed wall) acoustic cavity, random matrices $[\mathbf{D}]$ and $[\mathbf{K}]$ are positive and not invertible with rank $N - 1$, while for an almost closed (nonsealed wall) acoustic cavity, random matrices $[\mathbf{D}]$ and $[\mathbf{K}]$ are positive definite. The probability distributions of random matrices $[\mathbf{M}]$, $[\mathbf{D}]$, and $[\mathbf{K}]$ and of the $(N_S \times N)$ real random matrix $[\mathbf{C}]$ are constructed in Sections 9.8 to 9.10.

9.5 PRELIMINARY RESULTS FOR RANDOM MATRICES

In the framework of the nonparametric probabilistic approach of uncertainties, the probability distributions and the generators of independent realizations of such random matrices are constructed using random matrix theory (Mehta, 1991) and the maximum entropy principle (Jaynes, 1957; Soize, 2008) from information theory (Shannon, 1948), in which Shannon introduced the notion of entropy as a measure of the level of uncertainties for a probability distribution. For instance, if $p_X(x)$ is a probability density function on a real random variable X, the entropy $\mathcal{E}(p_X)$ of p_X is defined by $\mathcal{E}(p_X) = -\int_{-\infty}^{+\infty} p_X(x)\log(p_X(x))\,dx$. The maximum entropy principle consists in maximizing the entropy, that is to say, maximizing the uncertainties, under the constraints defined by the available information. Consequently, it is important to define the algebraic properties of the random matrices for which the probability distributions have to be constructed. Let E be the mathematical expectation which is such that $E\{X\} = \int_{-\infty}^{+\infty} x\,p_X(x)\,dx$. Consequently, we have $\mathcal{E}(p_X) = -E\{\log(p_X(X))\}$. In order to construct the probability distributions of the random matrices introduced in Section 9.4, we need to define a basic ensemble of random matrices.

It is well known that a real Gaussian random variable can take negative values. Consequently, the Gaussian orthogonal ensemble (GOE) of random matrices (Mehta, 1991) – which is the generalization for the matrix case of a random matrix for which the random entries are independent centered Gaussian random variables with same variances – cannot be used when positiveness property of the random matrix is required. Therefore, new ensembles of random matrices are required to implement the nonparametric probabilistic approach of uncertainties. In the following, we summarize the construction (Soize, 2000, 2001) of an ensemble of positive-definite symmetric ($m \times m$) real random matrices, which can be substituted by another one allowing a more flexible description of the dispersion levels to be introduced (Mignolet and Soize, 2008a).

9.5.1 Definition of the Available Information

For the probabilistic construction using the maximum entropy principle, the available information corresponds to two constraints. The first one is the mean value, which is given and equal to the identity matrix. The second one is an integrability condition, which has to be imposed in order to ensure the decreasing of the probability density function around the origin. These two constraints are written as

$$E\{[\mathbf{G}_0]\} = [I_m] \quad , \quad E\{\log(\det[\mathbf{G}_0])\} = \chi , \tag{9.7}$$

in which $|\chi|$ is finite and where $[I_m]$ is the $(m \times m)$ identity matrix.

9.5.2 Probability Density Function

The value of the probability density function of the random matrix $[\mathbf{G}_0]$ for the matrix $[G]$ is noted $p_{[\mathbf{G}_0]}([G])$ and satisfies the normalization condition

$$\int p_{[\mathbf{G}_0]}([G]) \, \widetilde{d}G = 1 , \tag{9.8}$$

in which the integration is carried out on the set of all the positive-definite symmetric $(m \times m)$ real matrices and where it can be shown that the volume element $\widetilde{d}G$ is written as $\widetilde{d}G = 2^{m(m-1)/4} \Pi_{1 \le j \le k \le m} \, dG_{jk}$.

Let δ be the positive real number defined by

$$\delta = \left\{ \frac{1}{m} E\{\| [\mathbf{G}_0] - [I_m] \|_F^2 \} \right\}^{1/2} , \tag{9.9}$$

which will allow the dispersion of random matrix $[\mathbf{G}_0]$ to be controlled and where $\| \mathcal{M} \|_F$ is the Frobenius matrix norm of the matrix $[\mathcal{M}]$ such that $\| \mathcal{M} \|_F^2 = \text{tr}\{[\mathcal{M}]^T[\mathcal{M}]\}$. For δ such that $0 < \delta < (m+1)^{1/2}(m+5)^{-1/2}$, the use of the maximum entropy principle under the two constraints defined by Eq. (9.7) and the normalization condition defined by Eq. (9.8), yields, for all positive-definite symmetric $(m \times m)$ real matrix $[G]$,

$$p_{[\mathbf{G}_0]}([G]) = c_0 \left(\det [G]\right)^{c_1} \exp\{-c_2 \, \text{tr}[G]\} , \tag{9.10}$$

in which the positive constant c_0, the constant $c_1 = (m+1)(1 - \delta^2)/(2\delta^2)$ and the constant $c_2 = (m+1)/(2\delta^2)$ depend on m and δ.

9.5.3 Generator of Independent Realizations

The generator of independent realizations of random matrix $[\mathbf{G}_0]$ whose probability density function is defined by Eq. (9.10) (which is required to solve the random equations with the Monte Carlo method) is constructed using the following algebraic representation. Using the Cholesky decomposition, random matrix $[\mathbf{G}_0]$ is written as $[\mathbf{G}_0] = [\mathbf{L}]^T [\mathbf{L}]$ in which $[\mathbf{L}]$ is an upper triangular $(m \times m)$ random matrix such that:

- Random variables $\{[\mathbf{L}]_{jj'}, j \leq j'\}$ are independent;
- For $j < j'$, the real-valued random variable $[\mathbf{L}]_{jj'}$ is written as $[\mathbf{L}]_{jj'} = \sigma_m U_{jj'}$ in which $\sigma_m = \delta(m+1)^{-1/2}$ and where $U_{jj'}$ is a real-valued Gaussian random variable with zero mean and variance equal to 1;
- For $j = j'$, the positive-valued random variable $[\mathbf{L}]_{jj}$ is written as $[\mathbf{L}]_{jj} = \sigma_m \sqrt{2V_j}$, in which V_j is a positive-valued Gamma random variable with probability density function $\Gamma(a_j, 1)$, in which $a_j = \frac{m+1}{2\delta^2} + \frac{1-j}{2}$.

9.5.4 Ensemble SG_ε^+ of Random Matrices

Let $0 \leq \varepsilon \ll 1$ be a positive number (for instance, ε can be chosen as 10^{-6}). We then define the ensemble SG_ε^+ of all the random matrices such that

$$[\mathbf{G}] = \frac{1}{1+\varepsilon}\{[\mathbf{G}_0] + \varepsilon [I_m]\}, \qquad (9.11)$$

in which $[\mathbf{G}_0]$ is a random matrix whose probability density function is defined in Section 9.5.2 and whose generator of independent realizations is defined in Section 9.5.3.

9.5.5 Case of Several Random Matrices

It can be proven that, if there are several random matrices for which there is no available information concerning their statistical dependencies, then the use of the maximum entropy principle yields that the best model that maximizes the entropy (the uncertainties) is a stochastic model for which all these random matrices are independent.

9.6 STOCHASTIC MODELING OF RANDOM MATRIX $[\mathbf{M}^S]$

Since there is no available information concerning the statistical dependency of $[\mathbf{M}^S]$ with the other random matrices of the problem, then random matrix $[\mathbf{M}^S]$ is independent of all the other random matrices. The deterministic matrix $[M^S]$ is positive definite and consequently can be written as $[M^S] = [L_{M^S}]^T [L_{M^S}]$ in which $[L_{M^S}]$ is an upper triangular real matrix. Using the nonparametric probabilistic approach of uncertainties, the stochastic model of the positive-definite symmetric random matrix $[\mathbf{M}^S]$ is then defined by

$$[\mathbf{M}^S] = [L_{M^S}]^T [\mathbf{G}_{M^S}] [L_{M^S}], \qquad (9.12)$$

where $[\mathbf{G}_{M^S}]$ is a $(N_S \times N_S)$ random matrix belonging to ensemble SG_ε^+ defined in Section 9.5.4 and whose probability distribution and generator of independent realizations depend only on dimension N_S and on the dispersion parameter δ_{M^S}.

9.7 STOCHASTIC MODELING OF $[\mathbf{D}^S(\omega)]$ AND $[\mathbf{K}^S(\omega)]$

Since there is no available information concerning the statistical dependency of the random matrices $\{[\mathbf{D}^S(\omega)], [\mathbf{K}^S(\omega)]\}$ with the other random matrices of the problem, then $\{[\mathbf{D}^S(\omega)], [\mathbf{K}^S(\omega)]\}$ are independent of all the other random matrices. But we will see below that $[\mathbf{D}^S(\omega)]$ and $[\mathbf{K}^S(\omega)]$ are statistically dependent random matrices. For stochastic modeling of $[\mathbf{D}^S(\omega)]$ and $[\mathbf{K}^S(\omega)]$, due to frequency-dependent linear constitutive equation of the mean computational vibroacoustic model (see Section 5.2), we propose to use the new extension presented in Soize and Poloskov (2012), which is based on the Hilbert transform (Papoulis, 1977) in the frequency range to express the causality properties (similarly to Section 5.2). The nonparametric probabilistic approach of uncertainties then consists in modeling the positive-definite symmetric ($N_S \times N_S$) real matrices $[D^S(\omega)]$ and $[K^S(\omega)]$ by random matrices $[\mathbf{D}^S(\omega)]$ and $[\mathbf{K}^S(\omega)]$ such that

$$E\{[\mathbf{D}^S(\omega)]\} = [D^S(\omega)], \ E\{[\mathbf{K}^S(\omega)]\} = [K^S(\omega)], \qquad (9.13)$$

$$[\mathbf{D}^S(-\omega)] = [\mathbf{D}(^S\omega)] , \quad [\mathbf{K}^S(-\omega)] = [\mathbf{K}^S(\omega)] . \qquad (9.14)$$

For $\omega \geq 0$, the construction of the stochastic model of the family of random matrices $[\mathbf{D}^S(\omega)]$ and $[\mathbf{K}^S(\omega)]$ is carried out as follows.

- Constructing the family $[\mathbf{D}^S(\omega)]$ of random matrices such that, for fixed ω, $[\mathbf{D}^S(\omega)] = [L_{D^s}(\omega)]^T [\mathbf{G}_{D^s}] [L_{D^s}(\omega)]$, where $[L_{D^s}(\omega)]$ is such that $[D^S(\omega)] = [L_{D^s}(\omega)]^T [L_{D^s}(\omega)]$ and where $[\mathbf{G}_{D^s}]$ is a ($N_S \times N_S$) random matrix belonging to ensemble SG_ε^+, defined in Section 9.5.4. Its probability distribution and its generator of independent realizations depend only on dimension N_S and on the dispersion parameter δ_{D^s}, which allows the level of uncertainties to be controlled.

- Constructing the random matrix $[\mathbf{K}^S(0)] = [L_{K^s(0)}]^T [\mathbf{G}_{K^s(0)}]$ $[L_{K^s(0)}]$ in which $[L_{K^s(0)}]$ is such that $[K^S(0)] = [L_{K^s(0)}]^T [L_{K^s(0)}]$ and where $[\mathbf{G}_{K^s(0)}]$ is a ($N_S \times N_S$) random matrix belonging to ensemble SG_ε^+ defined in Section 9.5.4. Its probability distribution and generator of independent realizations depend only on dimension N_S and on the dispersion parameter $\delta_{K^s(0)}$, which allows the level of uncertainties to be controlled. It should be noted that random matrix $[\mathbf{G}_{K^s(0)}]$ is independent of random matrix $[\mathbf{G}_{D^s}]$.

- For fixed $\omega \geq 0$, constructing the random matrix $[\mathbf{K}^S(\omega)]$ using the equation

$$[\mathbf{K}^S(\omega)] = [\mathbf{K}^S(0)] + \frac{\omega}{\pi} \,\text{p.v} \int_{-\infty}^{+\infty} \frac{1}{\omega - \omega'} [\mathbf{D}^S(\omega')] \, d\omega' , \quad (9.15)$$

or equivalently,

$$[\mathbf{K}^S(\omega)] = [\mathbf{K}^S(0)] + \frac{2\omega^2}{\pi} \,\text{p.v} \int_0^{+\infty} \frac{1}{\omega^2 - \omega'^2} [\mathbf{D}^S(\omega')] \, d\omega' , \quad (9.16)$$

which can also be rewritten as the following equation recommended for computation (because the singularity in $u = 1$ is independent of ω),

$$[\mathbf{K}^S(\omega)] = [\mathbf{K}^S(0)] + \frac{2\,\omega}{\pi}\,\text{p.v}\int_0^{+\infty} \frac{1}{1-u^2}\,[\mathbf{D}^S(\omega u)]\,du\,,$$

$$= [\mathbf{K}^S(0)] + \frac{2\,\omega}{\pi}\,\lim_{\eta\to 0}\{\int_0^{1-\eta} + \int_{1+\eta}^{+\infty}\}\,. \tag{9.17}$$

For fixed $\omega < 0$, $[\mathbf{K}^S(\omega)]$ is calculated using the even property, $[\mathbf{K}^S(\omega)] = [\mathbf{K}^S(-\omega)]$. With such a construction, it can be verified that, for all $\omega \geq 0$, $[\mathbf{K}^S(\omega)]$ is a positive-definite random matrix. In Soize and Poloskov (2012), the following sufficient condition is proven. If for all real vector $\mathbf{y} = (y_1, \ldots, y_{N_S})$, the random function $\mathbf{y}^T[\mathbf{D}(\omega)]\,\mathbf{y}$ is decreasing in ω for $\omega \geq 0$, then, for all $\omega \geq 0$, $[\mathbf{K}(\omega)]$ is a positive-definite random matrix.

If, for the whole structure Ω_S, the constitutive equation is modeled by an elastic material with a linear viscous damping term described in Section 5.2.2(i), that is to say, for all ω, $[\mathbf{D}^S(\omega)] = [\mathbf{D}^S(0)]$, then, from Eq. (9.17), it can easily be deduced that, for all ω, $[\mathbf{K}^S(\omega)] = [\mathbf{K}^S(0)]$.

9.8 STOCHASTIC MODELING OF [M]

Since there is no available information concerning the statistical dependency of $[\mathbf{M}]$ with the other random matrices of the problem, then random matrix $[\mathbf{M}]$ is independent of all the other random matrices. The deterministic matrix $[M]$ is positive definite and, consequently, can be written as $[M] = [L_M]^T\,[L_M]$ in which $[L_M]$ is an upper triangular real matrix. Using the nonparametric probabilistic approach of uncertainties, the stochastic model of the positive-definite symmetric random matrix $[\mathbf{M}]$ is then defined by

$$[\mathbf{M}] = [L_M]^T\,[\mathbf{G}_M]\,[L_M]\,, \tag{9.18}$$

where $[\mathbf{G}_M]$ is a $(N \times N)$ random matrix belonging to ensemble $\mathrm{SG}_\varepsilon^+$ defined in Section 9.5.4 and whose probability distribution and generator of independent realizations depend only on dimension N and on the dispersion parameter δ_M.

9.9 STOCHASTIC MODELING OF [D] AND [K]

Since there is no available information concerning the statistical dependency of $[\mathbf{D}]$ and $[\mathbf{K}]$ with the other random matrices of the problem, then random matrices $[\mathbf{D}]$ and $[\mathbf{K}]$ are mutually independent and are independent of all the other random matrices. For the stochastic modeling of $[\mathbf{D}]$ and $[\mathbf{K}]$, two cases have to be considered.

9.9.1 Closed (Sealed Wall) Acoustic Cavity

In such a case, the symmetric positive matrices $[D] = \tau\,[K]$ and $[K]$ are of rank $N - 1$ and can then be written as $[D] = [L_D]^T\,[L_D]$ and $[K] = [L_K]^T\,[L_K]$, in which $[L_K]$ is a rectangular $(N, N - 1)$ real matrix and where $[L_D] = \sqrt{\tau}\,[L_K]$. Using the nonparametric probabilistic approach of uncertainties, the stochastic models of the positive symmetric random matrices $[\mathbf{D}]$ and $[\mathbf{K}]$ of rank $N - 1$ are then defined by

$$[\mathbf{D}] = [L_D]^T\,[\mathbf{G}_D]\,[L_D] \quad , \quad [\mathbf{K}] = [L_K]^T\,[\mathbf{G}_K]\,[L_K], \quad (9.19)$$

where $[\mathbf{G}_D]$ and $[\mathbf{G}_K]$ are $((N - 1) \times (N - 1))$ independent random matrices belonging to ensemble $\mathrm{SG}_\varepsilon^+$ defined in Section 9.5.4 and whose probability distributions and generators of independent realizations depend only on dimension $N - 1$ and on the dispersion parameters δ_D and δ_K.

9.9.2 Almost Closed (Nonsealed Wall) Acoustic Cavity

The matrices $[D] = \tau\,[K]$ and $[K]$ are positive definite and thus invertible. Consequently, they can be written as $[D] = [L_D]^T\,[L_D]$ and

$[K] = [L_K]^T [L_K]$ in which $[L_K]$ is an upper triangular (N, N) real matrix and where $[L_D] = \sqrt{\tau} [L_K]$. Using the nonparametric probabilistic approach of uncertainties, the stochastic models of these positive symmetric random matrices yield

$$[\mathbf{D}] = [L_D]^T [\mathbf{G}_D][L_D] \quad , \quad [\mathbf{K}] = [L_K]^T [\mathbf{G}_K][L_K], \quad (9.20)$$

where $[\mathbf{G}_D]$ and $[\mathbf{G}_K]$ are $(N \times N)$ independent random matrices belonging to ensemble $\mathrm{SG}_\varepsilon^+$ defined in Section 9.5 and whose probability distributions and generators of independent realizations depend only on dimension N and on the dispersion parameters δ_D and δ_K.

9.10 STOCHASTIC MODELING OF [C]

Since there is no available information concerning the statistical dependency of [C] with the other random matrices of the problem, then random matrix [C] is independent of all the other random matrices. We use the construction proposed in Soize (2005b) in the context of the nonparametric probabilistic approach. Let us assumed that $N_S \geq N$ and that the $(N_S \times N)$ real matrix $[C]$ is such that $[C]\mathbf{q} = 0$ implies $\mathbf{q} = 0$. If $N \geq N_S$, the following construction must be applied to $[C]^T$ instead of $[C]$. Using the singular value decomposition of rectangular matrix $[C]$, one can write $[C] = [R][T]$ in which the $(N_S \times N)$ real matrix $[R]$ is such that $[R]^T [R] = [I_N]$ and where the symmetric square matrix $[T]$ is a positive-definite symmetric $(N \times N)$ real matrix. Using the Cholesky decomposition, we then have $[T] = [L_T]^T [L_T]$ in which $[L_T]$ is an upper triangular matrix. The $(N_S \times N)$ real random matrix [C] is then written as

$$[\mathbf{C}] = [R][\mathbf{T}] \quad , \quad [\mathbf{T}] = [L_T]^T [\mathbf{G}_C][L_T], \quad (9.21)$$

where $[\mathbf{G}_C]$ is a $(N \times N)$ random matrix belonging to ensemble $\mathrm{SG}_\varepsilon^+$ defined in Section 9.5.4 and whose probability distribution and generator of independent realizations depend only on dimension N_S and N, and on the dispersion parameter δ_C.

9.11 COMMENTS ABOUT STOCHASTIC
MODELS AND SOLVERS

The dispersion parameter δ of each random matrix $[\mathbf{G}]$ allows its level of dispersion (statistical fluctuations) to be controlled. The dispersion parameters of random matrices $[\mathbf{G}_{M^s}]$, $[\mathbf{G}_{D^s}]$, $[\mathbf{G}_{K^s(0)}]$, $[\mathbf{G}_M]$, $[\mathbf{G}_D]$, $[\mathbf{G}_K]$ and $[\mathbf{G}_C]$ are represented by a vector δ such that

$$\delta = (\delta_{M^s}, \delta_{D^s}, \delta_{K^s(0)}, \delta_M, \delta_D, \delta_K, \delta_C) \quad , \qquad (9.22)$$

which belongs to an admissible set $\mathcal{C}_\delta$ and which allows the level of uncertainties to be controlled for each type of operator introduced in the stochastic reduced-order computational vibroacoustic model. Consequently, if no experimental data are available, then δ has to be used to analyze the robustness of the solution of the vibroacoustic problem with respect to uncertainties by varying δ in $\mathcal{C}_\delta$.

For a given value of δ, there are two major classes of methods for solving the stochastic reduced-order computational vibroacoustic model defined by Eqs. (9.1) to (9.6). The first one belongs to the category of the spectral stochastic methods (see Ghanem and Spanos, 1991, 2003; Le Maître and Knio, 2010). The second one belongs to the class of stochastic sampling techniques for which the Monte Carlo method is the most popular. Such a method is often called nonintrusive since it offers the advantage of only requiring the availability of standard deterministic codes. It should be noted that the Monte Carlo numerical simulation method (see, for instance, Fishman, 1996; Rubinstein and Kroese, 2008) is a very effective and efficient one because it has the four following advantages:

- The method is non-intrusive,
- The method is adapted to massively parallel computation without any software developments,

- The convergence with respect to the number of realizations can be controlled during the computation,
- The speed of convergence is independent of the dimension.

If experimental data are available, then there are several possible methodologies (of which one is the maximum likelihood method) to identify the optimal values of δ. These aspects are not considered in this book and we refer the reader to Soize (2012a).

10

SYMMETRIC BEM WITHOUT SPURIOUS FREQUENCIES FOR THE EXTERNAL ACOUSTIC FLUID

The inviscid external acoustic fluid occupies infinite three-dimensional domain Ω_E, defined in Chapter 2, whose boundary $\partial\Omega_E$ is Γ_E (see Figure 10.1).

In the first part of this chapter, we present a construction of the frequency-dependent acoustic impedance boundary operator $\mathbf{Z}_{\Gamma_E}(\omega)$ introduced in Chapter 3, for the external acoustic problem, based on a symmetric boundary integral formulation without spurious frequencies.

In the second part, we present the finite element discretization of the frequency-dependent acoustic impedance boundary operator. We then obtain a symmetric *boundary element method* (BEM) without spurious frequencies for the external acoustic fluid.

In the third part, we express the pressure field $p_{\text{given}}|_{\Gamma_E}$ on Γ_E, introduced in Section 3.2 (Eq. (3.4)), and we construct the total pressure field $p_E(\mathbf{x}, \omega)$ in Ω_E.

It should be noted that many other computational methods can be found in literature for solving the exterior problem for the inviscid acoustic fluid. Among them, let us cite: (1) the artificial boundary conditions and the local/nonlocal nonreflecting boundary condition (NRBC) to take into account the Sommerfeld radiation condition at infinity, (2) the Dirichlet-to-Neumann (DtN) boundary condition related to a nonlocal artificial boundary condition, (3) the infinite element method, (4) the doubly asymptotic approximation method,

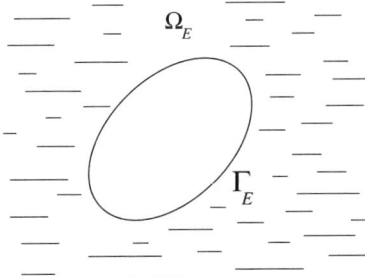

Figure 10.1 Geometry of the external infinite three-dimensional domain Ω_E with its boundary Γ_E.

(5) the finite element method in unbounded domain and related *a posteriori* error estimation and, finally, (6) the wave based method for unbounded domain; see, for instance, Geers and Felippa (1983); Givoli (1992); Harari et al. (1996); Astley (2000); Farhat et al. (2003, 2004); Oden et al. (2005); and Bergen et al. (2010).

The frequency-dependent acoustic impedance boundary operator $\mathbf{Z}_{\Gamma_E}(\omega)$ can be constructed either in the time domain with a Fourier transform post-processing or directly in the frequency domain. One technique consists in using integral equation methods (Jones, 1974; Roach, 1982; Costabel and Stephan, 1985; Jones, 1986; Kress, 1989; Colton and Kress, 1992; Dautray and Lions, 1992; Bonnet, 1999; Nedelec, 2001; Hsiao and Wendland, 2008).

In the time domain, the so-called Kirchhoff retarded potential formula is used (see, for instance, Baker and Copson, 1949; Lee et al., 2009).

On the other hand, in the frequency domain, the boundary element method can easily be implemented in massively parallel computers, which constitutes a significant advantage.

The finite element discretization of the boundary integral equation methods yields the boundary element methods (BEM) (Brebbia and Dominguez, 1992; Chen and Zhou, 1992; Hackbusch, 1995; Ohayon and Soize, 1998; Von Estorff et al., 2000; Gaul et al., 2003). Those formulations yield fully populated complex matrices. The computational cost can then be reduced using the fast multipole methods (Greengard and Rokhlin, 1987; Gumerov and Duraiswami, 2004; Schanz and Steinbach, 2007; Bonnet et al., 2009; Brunner et al., 2009).

Let us recall that, for the exterior problem related to the inviscid acoustic fluid (so called, in mathematics, *exterior Neumann problem related to the Helmholtz equation*), the boundary value problem has a unique solution for all real frequencies (Ohayon and Sanchez-Palencia, 1983; Sanchez-Hubert and Sanchez-Palencia, 1989; Dautray and Lions, 1992).

However, a major drawback of the boundary integral formulations commonly used for solving this exterior Neumann problem related to the Helmholtz equation is that the solution cannot be constructed for a sequence of real frequencies called *spurious* or *irregular frequencies*, also called *Jones eigenfrequencies* (Burton and Miller, 1971; Jones, 1983; Colton and Kress, 1992; Luke and Martin, 1995; Jentsch and Natroshvili, 1999). Various methods are proposed in the literature to overcome this mathematical difficulty arising in the boundary element methods (Panich, 1965; Schenck, 1968; Burton and Miller, 1971; Angelini and Hutin, 1983; Mathews, 1986; Amini and Harris, 1990; Amini et al., 1992; Ohayon and Soize, 1998). It should be noted that this drawback induces, at every spurious frequency and in its neighborhood, a spurious resonance in the frequency responses predicted by the computational vibroacoustic model. In particular, in the medium-frequency range there are usually a large number of such spurious frequencies and consequently, those corresponding spurious resonances cannot be identified in the predicted frequency responses, which are then erroneous.

In this chapter, we present a method that was initially developed by Angelini and Hutin (1983) in acoustics and which has been extended to electromagnetism (Angelini et al., 1993). This method is based on an appropriate symmetric boundary element method without spurious frequencies (i.e., valid for all real values of the frequency), which is numerically stable and very efficient. This method is detailed in

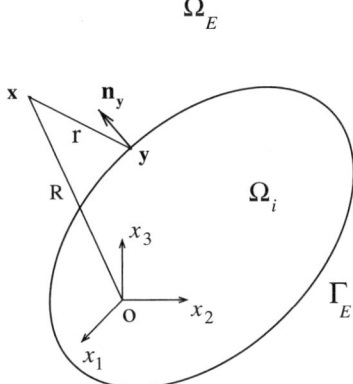

Figure 10.2 Geometry of the external infinite domain.

Ohayon and Soize (1998) and does not require the introduction of additional degrees of freedom in the numerical discretization for treatment of spurious frequencies.

10.1 EXTERIOR ACOUSTIC BOUNDARY VALUE PROBLEM

The geometry is defined in Figure 10.2. The external acoustic fluid occupies the infinite domain Ω_E.

For practical computational considerations, the exterior Neumann problem related to the Helmholtz equation is written in terms of a velocity potential $\psi(\mathbf{x}, \omega)$. Let $\mathbf{v}(\mathbf{x}, \omega) = \nabla \psi(\mathbf{x}, \omega)$ be the velocity field of the acoustic fluid. The acoustic pressure $p(\mathbf{x}, \omega)$ is related to $\psi(\mathbf{x}, \omega)$ by the following equation:

$$p(\mathbf{x}, \omega) = -i\omega \rho_E \psi(\mathbf{x}, \omega) \quad \text{in} \quad \Omega_E, \tag{10.1}$$

where ρ_E is the constant mass density of the acoustic fluid at equilibrium. Let c_E be the constant speed of sound in the acoustic fluid at equilibrium and let $k = \omega/c_E$ be the wave number at frequency ω. The exterior Neumann problem is written as

$$\nabla^2 \psi(\mathbf{x}, \omega) + k^2 \psi(\mathbf{x}, \omega) = 0 \quad \text{in} \quad \Omega_E, \qquad (10.2)$$

$$\frac{\partial \psi(\mathbf{y}, \omega)}{\partial \mathbf{n_y}} = v(\mathbf{y}) \quad \text{on} \quad \Gamma_E, \qquad (10.3)$$

$$|\psi| = O(\frac{1}{R}) \quad , \quad \left| \frac{\partial \psi}{\partial R} + i k \psi \right| = O(\frac{1}{R^2}), \qquad (10.4)$$

with $R = \|\mathbf{x}\| \to +\infty$, where $\partial/\partial R$ is the derivative in the radial direction and where $v(\mathbf{y})$ is the prescribed normal velocity field on Γ_E. Equation (10.2) is the Helmholtz equation in the external acoustic fluid, Eq. (10.3) is the Neumann condition on external fluid-structure interface Γ_E, and Eq. (10.4) corresponds to the outward Sommerfeld radiation condition at infinity, which means that if R goes to infinity, the quantity $|\psi|$ decreases in $1/R$ and $|\partial\psi/\partial R + i k \psi|$ decreases in $1/R^2$. These conditions show that only outward traveling waves are considered at infinity (outward traveling waves vanish at infinity and are not reflected) and that the radiated energy at infinity is lost.

The exterior Neumann problem related to the Helmholtz equation consists in finding, for all real ω, the solution ψ defined in $\Omega_E \cup \Gamma_E$, for prescribed v on Γ_E. It can be proven that this solution is unique for all real ω.

10.2 ACOUSTIC RADIATION AND BOUNDARY IMPEDANCE OPERATORS

For arbitrary real $\omega \neq 0$, it can be shown that the boundary value problem defined by Eqs. (10.2) to (10.4) admits a unique solution denoted ψ^{sol} that depends linearly of the normal velocity v (Sanchez-Hubert and Sanchez-Palencia, 1989; Dautray and Lions, 1992).

For $\mathbf{x}$ fixed in Ω_E, $\psi^{\text{sol}}(\mathbf{x}, \omega)$ can then be related to the prescribed normal velocity field v on Γ_E by introducing the linear operator

$\mathbf{R}(\mathbf{x}, \omega/c_E)$, such that

$$\psi^{\text{sol}}(\mathbf{x}, \omega) = \mathbf{R}(\mathbf{x}, \omega/c_E)\, v \,. \tag{10.5}$$

On the other hand, the value $\psi^{\text{sol}}_{\Gamma_E}$ of the field ψ^{sol} on Γ_E can then be related to the prescribed normal velocity field v on Γ_E by introducing the linear boundary operator $\mathbf{B}_{\Gamma_E}(\omega/c_E)$, such that

$$\psi^{\text{sol}}_{\Gamma_E} = \mathbf{B}_{\Gamma_E}(\omega/c_E)\, v \,. \tag{10.6}$$

- *Acoustic radiation impedance operator and pressure field in Ω_E.*
  Using Eqs. (10.1) and (10.5), for $\mathbf{x}$ fixed in Ω_E, the pressure $p(\mathbf{x}, \omega)$ can be written as

$$p(\mathbf{x}, \omega) = \mathbf{Z}_{\text{rad}}(\mathbf{x}, \omega)\, v \,, \tag{10.7}$$

  in which $\mathbf{Z}_{\text{rad}}(\mathbf{x}, \omega)$ is called the *acoustic radiation impedance operator*, which can then be written as

$$\mathbf{Z}_{\text{rad}}(\mathbf{x}, \omega) = -i\,\omega\,\rho_E\,\mathbf{R}(\mathbf{x}, \omega/c_E) \,. \tag{10.8}$$

  It should be noted that, for a fixed $\mathbf{x}$ in Ω_E, $\mathbf{Z}_{\text{rad}}(\mathbf{x}, \omega)$ (as well as $\mathbf{R}(\mathbf{x}, \omega/c_E)$) is a nonlocal operator, which means that the pressure $p(\mathbf{x}, \omega)$ depends on all the values of v on Γ_E and not only on the specific value $v(\mathbf{x})$ of v at $\mathbf{x}$.
- *Acoustic impedance boundary operator and pressure field on Γ_E.*
  Using Eqs. (10.1) and (10.6), the pressure field $p|_{\Gamma_E}(\omega)$ on Γ_E can be written as

$$p|_{\Gamma_E}(\omega) = \mathbf{Z}_{\Gamma_E}(\omega)\, v \,, \tag{10.9}$$

  in which $\mathbf{Z}_{\Gamma_E}(\omega)$ is called the *acoustic impedance boundary operator* and can then be written as

$$\mathbf{Z}_{\Gamma_E}(\omega) = -i\,\omega\,\rho_E\,\mathbf{B}_{\Gamma_E}(\omega/c_E) \,. \tag{10.10}$$

Let us note that $\mathbf{Z}_{\Gamma_E}(\omega)$ (as well as $\mathbf{B}_{\Gamma_E}(\omega/c_E)$) is a nonlocal operator, which means that the value $p|_{\Gamma_E}(\mathbf{x}, \omega)$ of the pressure field

$p|_{\Gamma_E}(\omega)$ at a fixed $\mathbf{x}$ on Γ_E, depends on all the values of v on Γ_E and not only on the specific value $v(\mathbf{x})$ of v at $\mathbf{x}$.

It should be noted that, since for all real ω the boundary value problem defined by Eqs. (10.2) to (10.4) has a unique solution, then $\mathbf{Z}_{\Gamma_E}(\omega)$ is unique for all real ω.

10.3 SYMMETRY AND POSITIVITY PROPERTIES OF THE BOUNDARY OPERATORS

The proof of the following properties can be found in Ohayon and Soize (1998).

- *Symmetry property of $\mathbf{Z}_{\Gamma_E}(\omega)$.* The transpose of operator $\mathbf{B}_{\Gamma_E}(\omega/c_E)$ is denoted by $^t\mathbf{B}_{\Gamma_E}(\omega/c_E)$. The following symmetry property can then be proved:

$$^t\mathbf{B}_{\Gamma_E}(\omega/c_E) = \mathbf{B}_{\Gamma_E}(\omega/c_E). \qquad (10.11)$$

From Eqs. (10.10) and (10.11), the symmetry property for the acoustic impedance boundary operator can be deduced,

$$^t\mathbf{Z}_{\Gamma_E}(\omega) = \mathbf{Z}_{\Gamma_E}(\omega). \qquad (10.12)$$

It should be noted that these complex operators are symmetric but not hermitian.

- *Positivity of the real part of $\mathbf{Z}_{\Gamma_E}(\omega)$.* Let $\Re e\, \mathbf{Z}_{\Gamma_E}(\omega)$ and $\Im m\, \mathbf{Z}_{\Gamma_E}(\omega)$ be the real and the imaginary parts of the complex operator $\mathbf{Z}_{\Gamma_E}(\omega)$. Operator $i\omega\mathbf{Z}_{\Gamma_E}(\omega)$ can be written as

$$i\omega\mathbf{Z}_{\Gamma_E}(\omega) = -\omega^2\,\mathbf{M}_{\Gamma_E}(\omega/c_E) + i\omega\mathbf{D}_{\Gamma_E}(\omega/c_E), \qquad (10.13)$$

in which $\mathbf{M}_{\Gamma_E}(\omega/c_E)$ and $\mathbf{D}_{\Gamma_E}(\omega/c_E)$ are two linear operators such that

$$\omega\,\mathbf{M}_{\Gamma_E}(\omega/c_E) = \Im m\, \mathbf{Z}_{\Gamma_E}(\omega), \qquad (10.14)$$

$$\mathbf{D}_{\Gamma_E}(\omega/c_E) = \Re e\, \mathbf{Z}_{\Gamma_E}(\omega). \qquad (10.15)$$

Due to the Sommerfeld radiation condition at infinity, it can be shown that the symmetric real operator $\mathbf{D}_{\Gamma_E}(\omega/c_E)$ is positive.

10.4 CONSTRUCTION OF THE ACOUSTIC IMPEDANCE BOUNDARY OPERATOR

In this section, we present the appropriate *symmetric boundary formulation without spurious frequencies*, for which details can be found in Ohayon and Soize (1998). This formulation simultaneously uses two boundary integral equations on Γ_E. The first one is the classical integral representation on Γ_E for the Helmholtz equation in the external infinite three-dimensional domain Ω_E. The second integral equation is obtained by the normal derivative on Γ_E of the first one. We then construct the following appropriate formulation:

$$
\begin{bmatrix} 0 \\ \psi^{sol}_{\Gamma_E} \end{bmatrix} = \begin{bmatrix} -\mathbf{S}_T(\omega/c_E) & \frac{1}{2}\,{}^t\mathbf{I} - {}^t\mathbf{S}_D(\omega/c_E) \\ \frac{1}{2}\,\mathbf{I} - \mathbf{S}_D(\omega/c_E) & \mathbf{S}_S(\omega/c_E) \end{bmatrix} \begin{bmatrix} \psi_{\Gamma_E} \\ v \end{bmatrix}. \quad (10.16)
$$

In Eq. (10.16), v is given and we have to calculate $\psi^{sol}_{\Gamma_E}$ in eliminating the potential field ψ_{Γ_E} defined on Γ_E between the two block equations (the role played by ψ_{Γ_E} and its elimination are explained in the following). More precisely, symmetric Eq. (10.16) allows $\psi^{sol}_{\Gamma_E}$ to be calculated as a function of v for all real ω (without effects of spurious frequencies, as explained in the following) and, consequently, allows the operator $\mathbf{B}_{\Gamma_E}(\omega/c_E)$, which relates $\psi^{sol}_{\Gamma_E}$ to v (see Eq. (10.6)), to be constructed. In Eq. (10.16), the linear boundary integral operators $\mathbf{S}_S(\omega/c_E)$, $\mathbf{S}_D(\omega/c_E)$, and $\mathbf{S}_T(\omega/c_E)$ are defined by

$$
<\mathbf{S}_S(\omega/c_E)\,v\,,\,\delta v> = \int_{\Gamma_E} \int_{\Gamma_E} G(\mathbf{x}-\mathbf{y})\,v(\mathbf{y})\,\delta v(\mathbf{x})\,ds_\mathbf{y}\,ds_\mathbf{x}\,, \quad (10.17)
$$

$$
<\mathbf{S}_D(\omega/c_E)\,\psi_{\Gamma_E}\,,\,\delta v> = \int_{\Gamma_E} \int_{\Gamma_E} \frac{\partial G(\mathbf{x}-\mathbf{y})}{\partial \mathbf{n_y}}\,\psi_{\Gamma_E}(\mathbf{y})\,\delta v(\mathbf{x})\,ds_\mathbf{y}\,ds_\mathbf{x}\,, \quad (10.18)
$$

$$<\mathbf{S}_T(\omega/c_E)\psi_{\Gamma_E}, \delta\psi_{\Gamma_E}> = -k^2\!\!\int_{\Gamma_E}\!\int_{\Gamma_E} G(\mathbf{x}-\mathbf{y})\,\mathbf{n_x}\cdot\mathbf{n_y}\,\psi_{\Gamma_E}(\mathbf{y})\,\delta\psi_{\Gamma_E}(\mathbf{x})\,ds_\mathbf{y}\,ds_\mathbf{x}$$

$$+\int_{\Gamma_E}\!\int_{\Gamma_E} G(\mathbf{x}-\mathbf{y})\,\{\mathbf{n_y}\times\nabla_\mathbf{y}\psi_{\Gamma_E}(\mathbf{y})\}\cdot\{\mathbf{n_x}\times\nabla_\mathbf{x}\delta\psi_{\Gamma_E}(\mathbf{x})\}\,ds_\mathbf{y}\,ds_\mathbf{x}\,.$$

$$(10.19)$$

where $G(\mathbf{x}-\mathbf{y})$ is the Green function, which is written as

$$G(\mathbf{x}-\mathbf{y}) = -(4\pi)^{-1}\,e^{-ikr}/r\,, \qquad (10.20)$$

with $r = \|\mathbf{x}-\mathbf{y}\|$ and where $\mathbf{I}$ is the identity operator. In Eqs. (10.17) to (10.19), the brackets correspond to bilinear forms that allow the operators to be defined and the functions δv and $\delta\psi_{\Gamma_E}$ are associated with functions v and ψ_{Γ_E}. The cross symbol $\times$ denotes the vector product and the dot denotes the scalar product in the three-dimensional Euclidean space. Let $\mathbf{H}(\omega/c_E)$ be the operator introduced in Eq. (10.16) and defined by

$$\mathbf{H}(\omega/c_E) = \begin{bmatrix} -\mathbf{S}_T(\omega/c_E) & \frac{1}{2}\,{}^t\mathbf{I} - {}^t\mathbf{S}_D(\omega/c_E) \\ \frac{1}{2}\mathbf{I} - \mathbf{S}_D(\omega/c_E) & \mathbf{S}_S(\omega/c_E) \end{bmatrix}. \qquad (10.21)$$

It can be shown that operator $\mathbf{H}(\omega/c_E)$ is symmetric, ${}^t\mathbf{H}(\omega/c_E) = \mathbf{H}(\omega/c_E)$. In Eq. (10.16), the first equation can be rewritten as $\mathbf{S}_T(\omega/c_E)\,\psi_{\Gamma_E} = (\frac{1}{2}\,{}^t\mathbf{I} - {}^t\mathbf{S}_D(\omega/c_E))\,v$. This boundary equation which allows the velocity potential to be calculated for a given normal velocity, has a unique solution for all real ω that does not belong to the set of frequencies for which $\mathbf{S}_T(\omega/c_E)$ has a null space that is not reduced to $\{0\}$. This set of frequencies is called the set of the *spurious* or *irregular* or *Jones* frequencies. Consequently, as proven in Ohayon and Soize (1998), for a spurious frequency, ψ_{Γ_E} is the sum of solution $\psi_{\Gamma_E}^{sol}$ with an arbitrary element belonging to the null space of operator $\mathbf{S}_T(\omega/c_E)$. The originality of the proposed method (Angelini and Hutin, 1983; Ohayon and Soize, 1998) then consists in using the second equation, which is written as $\psi_{\Gamma_E}^{sol} = (\frac{1}{2}\mathbf{I} - \mathbf{S}_D(\omega/c_E))\,\psi_{\Gamma_E} + \mathbf{S}_S(\omega/c_E)\,v$, and which yields solution $\psi_{\Gamma_E}^{sol}$ for all real ω, because the elements belonging to the null space are filtered when ω is a spurious frequency. Concerning the practical construction of $\psi_{\Gamma_E}^{\text{sol}}$, for all real values of ω, using Eq. (10.16), a particular elimination procedure will be described in Section 10.7.

10.5 CONSTRUCTION OF THE ACOUSTIC RADIATION IMPEDANCE OPERATOR

The solution $\{\psi^{\text{sol}}(\mathbf{x}, \omega), \mathbf{x} \in \Omega_E\}$ of Eqs. (10.2) to (10.4) can be calculated using the following integral equation:

$$\psi^{\text{sol}}(\mathbf{x}, \omega) = \int_{\Gamma_E} \{G(\mathbf{x} - \mathbf{y})\, v(\mathbf{y}) - \psi_{\Gamma_E}^{\text{sol}}(\mathbf{y}, \omega)\, \frac{\partial G(\mathbf{x} - \mathbf{y})}{\partial \mathbf{n_y}}\}\, ds_{\mathbf{y}}. \quad (10.22)$$

For all $\mathbf{x}$ fixed in Ω_E, we define the linear integral operators $\mathbf{R}_S(\mathbf{x}, \omega/c_E)$ and $\mathbf{R}_D(\mathbf{x}, \omega/c_E)$ by

$$\mathbf{R}_S(\mathbf{x}, \omega/c_E)\, v = \int_{\Gamma_E} G(\mathbf{x} - \mathbf{y})\, v(\mathbf{y})\, ds_{\mathbf{y}}, \quad (10.23)$$

$$\mathbf{R}_D(\mathbf{x}, \omega/c_E)\, \psi_{\Gamma_E} = \int_{\Gamma_E} \psi_{\Gamma_E}(\mathbf{y})\, \frac{\partial G(\mathbf{x} - \mathbf{y})}{\partial \mathbf{n_y}}\, ds_{\mathbf{y}}. \quad (10.24)$$

Using Eq. (10.6), Eq. (10.22) can be rewritten as

$$\psi^{\text{sol}}(\mathbf{x}, \omega) = \{\mathbf{R}_S(\mathbf{x}, \omega/c_E) - \mathbf{R}_D(\mathbf{x}, \omega/c_E)\, \mathbf{B}_{\Gamma_E}(\omega/c_E)\}\, v. \quad (10.25)$$

From Eq. (10.5), we deduce that, for all $\mathbf{x}$ fixed in Ω_E,

$$\mathbf{R}(\mathbf{x}, \omega/c_E) = \mathbf{R}_S(\mathbf{x}, \omega/c_E) - \mathbf{R}_D(\mathbf{x}, \omega/c_E)\, \mathbf{B}_{\Gamma_E}(\omega/c_E), \quad (10.26)$$

and the acoustic radiation impedance operator $\mathbf{Z}_{\text{rad}}(\mathbf{x}, \omega)$ is calculated using Eqs. (10.8) and (10.26),

$$\mathbf{Z}_{\text{rad}}(\mathbf{x}, \omega) = -i\omega \rho_E \{\mathbf{R}_S(\mathbf{x}, \omega/c_E) - \mathbf{R}_D(\mathbf{x}, \omega/c_E)\, \mathbf{B}_{\Gamma_E}(\omega/c_E)\}. \quad (10.27)$$

10.6 SYMMETRIC BOUNDARY ELEMENT METHOD WITHOUT SPURIOUS FREQUENCIES

The boundary element method (BEM) consists in using the finite element method in order to discretize the boundary integral operators

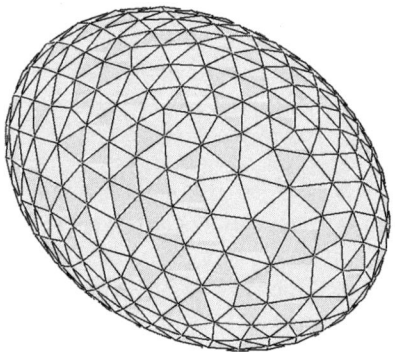

Figure 10.3 Finite element mesh of boundary Γ_E.

$\mathbf{S}_S(\omega/c_E)$, $\mathbf{S}_D(\omega/c_E)$, and $\mathbf{S}_T(\omega/c_E)$. Let us consider a finite element mesh of boundary Γ_E (see Figure 10.3).

Let $\mathbf{V} = (V_1, \ldots, V_{n_E})$ and $\mathbf{\Psi}_{\Gamma_E} = (\Psi_{\Gamma_E,1}, \ldots, \Psi_{\Gamma_E,n_E})$ be the complex vectors of the n_E degrees-of-freedom constituted of the values of v and ψ_{Γ_E} at the nodes of the mesh. Let $[S_S(\omega/c_E)]$, $[S_D(\omega/c_E)]$ and $[S_T(\omega/c_E)]$ be the full complex matrices corresponding to the discretization of the operators defined in Eqs. (10.17) to (10.19). The complex matrices $[S_S(\omega/c_E)]$ and $[S_T(\omega/c_E)]$ are symmetric. The finite element discretization of Eq. (10.16) yields

$$\begin{bmatrix} \mathbf{0} \\ \mathbf{\Psi}_{\Gamma_E}^{\text{sol}} \end{bmatrix} = [H(\omega/c_E)] \begin{bmatrix} \mathbf{\Psi}_{\Gamma_E} \\ \mathbf{V} \end{bmatrix}, \qquad (10.28)$$

in which the symmetric complex matrix $[H(\omega/c_E)]$ is written as

$$\begin{bmatrix} -[S_T(\omega/c_E)] & \frac{1}{2}[E]^T - [S_D(\omega/c_E)]^T \\ \frac{1}{2}[E] - [S_D(\omega/c_E)] & [S_S(\omega/c_E)] \end{bmatrix}. \qquad (10.29)$$

In Eq. (10.28), $\mathbf{\Psi}_{\Gamma_E}^{\text{sol}}$ is the complex vector of the nodal unknowns corresponding to the finite element discretization of $\psi_{\Gamma_E}^{\text{sol}}$. The matrix $[E]$ is the nondiagonal $(n_E \times n_E)$ real matrix corresponding to the discretization of identity operator $\mathbf{I}$. The elimination of $\mathbf{\Psi}_{\Gamma_E}$ in Eq. (10.28) yields a linear equation between $\mathbf{\Psi}_{\Gamma_E}^{\mathrm{sol}}$ and $\mathbf{V}$, which defines the symmetric $(n_E \times n_E)$ complex matrix $[B_{\Gamma_E}(\omega/c_E)]$ that corresponds to the finite element discretization of boundary integral operator $\mathbf{B}_{\Gamma_E}(\omega/c_E)$. We then have

$$\mathbf{\Psi}_{\Gamma_E}^{\mathrm{sol}} = [B_{\Gamma_E}(\omega/c_E)]\, \mathbf{V}. \tag{10.30}$$

Numerical method for eliminating $\mathbf{\Psi}_{\Gamma_E}$. The particular elimination procedure discussed in Section 10.4 is defined in the following. Vector $\mathbf{\Psi}_{\Gamma_E}$ is eliminated using a Gauss elimination with a partial row pivoting algorithm (Golub and Van Loan, 1989). If ω does not belong to the set of the spurious frequencies, then $[S_T(\omega/c_E)]$ is invertible and the elimination in Eq. (10.28) is performed up to row number n_E. If ω coincides with a spurious frequency ω_α, that is to say, $\omega = \omega_\alpha$, then $[S_T(\omega_\alpha/c_E)]$ is not invertible and its null space is a real subspace of $\mathbb{C}^{n_E}$ of dimension $n_\alpha < n_E$. In this case, the elimination in Eq. (10.28) is performed up to row number $n_E - n_\alpha$. In practice, n_α is unknown. During the Gauss elimination with a partial row pivoting algorithm, the elimination process is stopped when a "zero" pivot is encountered. It should be noted that when the elimination is stopped, the equations corresponding to row numbers $n_E - n_\alpha + 1, \ldots, n_E$ are automatically satisfied.

Construction of the acoustic impedance boundary matrix $[Z_{\Gamma_E}(\omega)]$. From Eq. (10.10), we deduce that the $(n_E \times n_E)$ complex symmetric matrix $[Z_{\Gamma_E}(\omega)]$ of the acoustic impedance boundary operator $\mathbf{Z}_{\Gamma_E}(\omega)$ is written as

$$[Z_{\Gamma_E}(\omega)] = -i\,\omega\,\rho_E\,[B_{\Gamma_E}(\omega/c_E)]. \tag{10.31}$$

Let $\mathfrak{Re}\,[Z_{\Gamma_E}(\omega)]$ and $\mathfrak{Im}\,[Z_{\Gamma_E}(\omega)]$ be the real and the imaginary parts of the complex matrix $[Z_{\Gamma_E}(\omega)]$. Complex matrix $i\omega\,[Z_{\Gamma_E}(\omega)]$ can be written as

$$i\omega\,[Z_{\Gamma_E}(\omega)] = -\omega^2\,[M_{\Gamma_E}(\omega/c_E)] + i\omega\,[D_{\Gamma_E}(\omega/c_E)], \tag{10.32}$$

in which $[M_{\Gamma_E}(\omega/c_E)]$ and $[D_{\Gamma_E}(\omega/c_E)]$ are two symmetric real matrices such that

$$\omega\,[M_{\Gamma_E}(\omega/c_E)] = \Im m\,[Z_{\Gamma_E}(\omega)]\,, \tag{10.33}$$

$$[D_{\Gamma_E}(\omega/c_E)] = \Re e\,[Z_{\Gamma_E}(\omega)]\,. \tag{10.34}$$

Due to the Sommerfeld radiation condition at infinity, it can be shown that the symmetric real matrix $[D_{\Gamma_E}(\omega/c_E)]$ is positive.

Construction of the acoustic radiation impedance matrix $[Z_{rad}(\mathbf{x}, \omega)]$. Let $[R_S(\mathbf{x}, \omega/c_E)]$ and $[R_D(\mathbf{x}, \omega/c_E)]$ be the complex matrices corresponding to the finite element discretization of the operators $\mathbf{R}_S(\mathbf{x}, \omega/c_E)$ and $\mathbf{R}_D(\mathbf{x}, \omega/c_E)$ defined by Eqs. (10.23) and (10.24). Consequently, using the previous construction of matrix $[B_{\Gamma_E}(\omega/c_E)]$, the finite element discretization of the acoustic radiation impedance operator $\mathbf{Z}_{rad}(\mathbf{x}, \omega)$ defined by Eq. (10.27) is written as

$$[Z_{rad}(\mathbf{x}, \omega)] = -i\,\omega\,\rho_E\,\{[R_S(\mathbf{x}, \omega/c_E)] - [R_D(\mathbf{x}, \omega/c_E)]\,[B_{\Gamma_E}(\omega/c_E)]\}\,. \tag{10.35}$$

10.7 PRESSURE FIELD INDUCED BY THE EXTERNAL ACOUSTIC SOURCE

Expression of the pressure $p_{given}(\mathbf{x}, \omega)$ *for* $\mathbf{x}$ *in* Ω_E *induced by the external acoustic source density.* For $\mathbf{x}$ fixed in Ω_E, we introduce the pressure $p_{given}(\mathbf{x}, \omega)$, defined as the superposition of the incident pressure $p_{inc,Q}(\mathbf{x}, \omega)$ induced by the external acoustic source density $Q_E(\mathbf{y}, \omega)$ (introduced in Chapter 3) with the pressure $p_{rig}(\mathbf{x}, \omega)$ induced by the scattering of the incident wave $p_{inc,Q}$ by the boundary Γ_E considered as a rigid wall. It can then be shown that

$$p_{given}(\mathbf{x}, \omega) = p_{inc,Q}(\mathbf{x}, \omega) - \mathbf{Z}_{rad}(\mathbf{x}, \omega)\{\frac{\partial\psi_{inc,Q}}{\partial\mathbf{n_y}}\}\,, \tag{10.36}$$

where $p_{inc,Q}(\mathbf{x}, \omega)$ is the pressure in the free space induced by the acoustic source Q_E, which is written as

$$p_{inc,Q}(\mathbf{x}, \omega) = -i\omega \int_{K_Q} G(\mathbf{x} - \mathbf{y})\,Q_E(\mathbf{y}, \omega)\,d\mathbf{y}\,, \tag{10.37}$$

in which the Green function G is defined by Eq. (10.20) and where

$$\frac{\partial \psi_{\text{inc},Q}}{\partial \mathbf{n_y}}(\mathbf{y}, \omega) = \frac{1}{\rho_E} \int_{K_Q} \frac{\partial G(\mathbf{y} - \mathbf{y}')}{\partial \mathbf{n_y}} Q_E(\mathbf{y}', \omega) \, d\mathbf{y}', \quad (10.38)$$

for which Eq. (10.1) has been used.

Expression of the pressure field $p_{\text{given}}|_{\Gamma_E}$ *on* Γ_E *induced by the external acoustic source density.* The pressure field $p_{\text{given}}|_{\Gamma_E}$ introduced in Section 3.2 is defined as the value of the pressure field p_{given} on Γ_E, which is then written as

$$p_{\text{given}}|_{\Gamma_E}(\omega) = p_{\text{inc},Q}|_{\Gamma_E}(\omega) - Z_{\Gamma_E}(\omega)\{\frac{\partial \psi_{\text{inc},Q}}{\partial \mathbf{n_y}}\}, \quad (10.39)$$

in which $p_{\text{inc},Q}|_{\Gamma_E}(\omega)$ is defined by Eq. (10.37) taking $\mathbf{x}$ in Γ_E and where $\partial \psi_{\text{inc},Q}/\partial \mathbf{n_y}$ is defined by Eq. (10.38) taking $\mathbf{y}$ in Γ_E.

Finite element discretization of $p_{\text{given}}|_{\Gamma_E}$ *on* Γ_E *and* $p_{\text{given}}(\mathbf{x}, \omega)$ *for* $\mathbf{x}$ *in* Ω_E. Let $\mathbf{V}_{\text{inc},Q}$ be the complex vectors of the n_E degrees of freedom corresponding to the finite element discretization of the field $\partial \psi_{\text{inc},Q}/\partial \mathbf{n_y}$ on Γ_E. Let $\mathbf{P}_{\text{inc},Q}|_{\Gamma_E}$ and $\mathbf{P}_{\text{given}}|_{\Gamma_E}$ be the complex vectors given by the finite element discretization of the fields $p_{\text{inc},Q}|_{\Gamma_E}$ and $p_{\text{given}}|_{\Gamma_E}$ and constituted of the values of these fields at the nodes of the finite element mesh of Γ_E.

The finite element discretization of Eq. (10.39) can then written as

$$\mathbf{P}_{\text{given}}|_{\Gamma_E}(\omega) = \mathbf{P}_{\text{inc},Q}|_{\Gamma_E}(\omega) - [Z_{\Gamma_E}(\omega)]\mathbf{V}_{\text{inc},Q}(\omega). \quad (10.40)$$

10.8 RESULTANT PRESSURE FIELD IN THE EXTERNAL DOMAIN

Resultant pressure $p_E(\mathbf{x}, \omega)$ *in* Ω_E. For $\mathbf{x}$ fixed in Ω_E and due to the linearity of the problem, the resultant pressure $p_E(\mathbf{x}, \omega)$ can be written as the superposition of (i) $p_{\text{given}}(\mathbf{x}, \omega)$ given by Eq. (10.36) (which is itself the superposition of the incident pressure $p_{\text{inc},Q}(\mathbf{x}, \omega)$ induced by $Q_E(\mathbf{y}, \omega)$, with the pressure $p_{\text{rig}}(\mathbf{x}, \omega)$ induced by the scattering of the incident wave $p_{\text{inc},Q}$ by the boundary Γ_E considered as a rigid wall) with (ii) the radiation field $p_{\text{rad}}(\mathbf{x}, \omega)$ induced by the deformation of boundary Γ_E represented by the normal displacement field $\mathbf{u}(\omega) \cdot \mathbf{n}$.

We can then write,

$$p_E(\mathbf{x}, \omega) = p_{\text{given}}(\mathbf{x}, \omega) + p_{\text{rad}}(\mathbf{x}, \omega), \qquad (10.41)$$

in which $p_{\text{rad}}(\mathbf{x}, \omega)$ is given by

$$p_{\text{rad}}(\mathbf{x}, \omega) = i\omega\, \mathbf{Z}_{\text{rad}}(\mathbf{x}, \omega)\{\mathbf{u}(\omega) \cdot \mathbf{n}\}. \qquad (10.42)$$

Finite element approximation of the resultant pressure $p_E(x, \omega)$ in Ω_E. Let $\mathbf{U_n} = (U_{\mathbf{n},1}, \ldots, U_{\mathbf{n},n_E})$ be the complex vector of the n_E degrees of freedom constituted of the values of $\mathbf{u}(\omega) \cdot \mathbf{n}$ at the nodes of the mesh of Γ_E. Therefore, the finite element approximation of Eq. (10.41) is written as

$$p_E(\mathbf{x}, \omega) \simeq p_{\text{inc,Q}}(\mathbf{x}, \omega) - [Z_{\text{rad}}(\mathbf{x}, \omega)]\,\mathbf{V}_{\text{inc,Q}}(\omega) + i\omega\,[Z_{\text{rad}}(\mathbf{x}, \omega)]\,\mathbf{U_n}(\omega),$$
$$(10.43)$$

in which the complex matrix $[Z_{\text{rad}}(\mathbf{x}, \omega)]$ is defined by Eq. (10.35).

REFERENCES

Amini, S., and Harris, P. J. 1990. A comparison between various boundary integral formulations of the exterior acoustic problem. *Computer Methods in Applied Mechanics and Engineering*, **84**, 59–75.

Amini, S., Harris, P. J., and Wilton, D. T. 1992. *Coupled Boundary and Finite Element Methods for the Solution of the Dynamic Fluid-Structure Interaction Problem*. Lecture Notes in Eng., Vol 77. New York: Springer.

Andrianarison, O., and Ohayon, R. 2006. Reduced models for modal analysis of fluid-structure systems taking into account compressibility and gravity effects. *Computer Methods in Applied Mechanics and Engineering*, **195**, 5656–5672.

Angelini, J. J., and Hutin, P. M. 1983. Exterior Neumann problem for Helmholtz equation. Problem of irregular frequencies. *La Recherche Aérospatiale*, **3**, 43–52 (English edition).

Angelini, J. J., Soize, C., and Soudais, P. 1993. Hybrid numerical method for harmonic 3D Maxwell equations: scattering by mixed conducting and inhomogeneous-anisotropic dielectric media. *IEEE Transaction on Antennas and Propagation*, **41**(1), 66–76.

Arnst, M., Clouteau, D., Chebli, H., Othman, R., and Degrande, G. 2006. A nonparametric probabilistic model for ground-borne vibrations in buildings. *Probabilistic Engineering Mechanics*, **21**(1), 18–34.

Astley, R. J. 2000. Infinite elements for wave problems: a review of current formulations and assessment of accuracy. *International Journal of Numerical Methods in Engineering*, **49**(7), 951–976.

Bagley, R., and Torvik, P. 1983. Fractional calculus: a different approach to the analysis of viscoelastically damped struture. *AIAA Journal*, **5**(5), 741–748.

Baker, B. B., and Copson, E. T. 1949. *The Mathematical Theory of Huygens Principle*. Oxford: Clarendon Press.

Batchelor, G. K. 2000. *An Introduction to Fluid Dynamics*. Cambridge: Cambridge University Press.

Bathe, K. J., and Wilson, E. L. 1976. *Numerical Methods in Finite Element Analysis*. New York: Prentice-Hall.

Batou, A., and Soize, C. 2009a. Experimental identification of turbulent fluid forces applied to fuel assemblies using an uncertain model and fretting-wear estimation. *Mechanical Systems and Signal Processing*, **23**(7), 2141–2153.

Batou, A., and Soize, C. 2009b. Identification of stochastic loads applied to a non-linear dynamical system using an uncertain computational model and experimental responses. *Computational Mechanics*, **43**(4), 559–571.

Batou, A., Soize, C., and Corus, M. 2011. Experimental identification of an uncertain computational dynamical model representing a family of structures. *Computer and Structures*, **89**(13–14), 1440–1448.

Bazilevs, Y., Takizawa, K., and Tezduyar, T. E. 2013. *Computational Fluid-Structure Interaction*. Wiley Series in Computational Mechanics. Chichester: John Wiley & Sons.

Beck, J. L., and Au, S. K. 2002. Bayesian updating of structural models and reliability using Markov chain Monte Carlo simulation. *Journal of Engineering Mechanics – ASCE*, **128**(4), 380–391.

Beck, J. L., and Katafygiotis, L. S. 1998. Updating models and their uncertainties. I: Bayesian statistical framework. *Journal of Engineering Mechanics – ASCE*, **124**(4), 455–461.

Beltman, W. M. 1999. Viscothermal wave propagation including acousto-elastic interaction, Part I: Theory. Part II: Application. *Journal of Sound and Vibration*, **227**(3), 555–586, 587–609.

Bergen, B., Van Genechten, B., Vandepitte, D., and Desmet, W. 2010. An efficient Trefftz-based method for three-dimensional Helmholtz in unbounded domain. *Computer Modeling in Engineering & Sciences*, **61**(2), 155–175.

Blackstock, D. T. 2000. *Fundamentals in Physical Acoustics*. New York: John Wiley & Sons.

Bland, D. R. 1960. *The Theory of Linear Viscoelasticity*. London: Pergamon.

Bonnet, M. 1999. *Boundary Integral Equation Methods for Solids and Fluids*. New York: John Wiley & Sons.

Bonnet, M., Chaillat, S., and Semblat, J. F. 2009. Multi-level fast multipole BEM for 3-D elastodynamics. Pages 15–27 of: Manomis, G. D., and Polyzos, D. (eds), *Recent Advances in Boundary Element Methods*. Heidelberg: Springer.

Brebbia, C. A., and Dominguez, J. 1992. *Boundary Elements: An Introductory Course*. New York: McGraw-Hill.

Bruneau, M. 2006. *Fundamentals of Acoustics*. Newport Beach: ISTE USA.

Brunner, D., Junge, M., and Gaul, L. 2009. A comparison of FE-BE coupling schemes for large scale problems with fluid-structure interaction. *International Journal of Numerical Methods in Engineering*, **77**(5), 664–688.

Burton, A. J., and Miller, G. F. 1971. The application of integral equation methods to the numerical solution of some exterior boundary value problems. *Proceeding of the Royal Society of London. Series A*, **323**, 201–210.

Capiez-Lernout, E., and Soize, C. 2004. Nonparametric modeling of random uncertainties for dynamic response of mistuned bladed disks. *Journal of Engineering for Gas Turbines and Power*, **126**(3), 600–618.

Capiez-Lernout, E., and Soize, C. 2008a. Design optimization with an uncertain vibroacoustic model. *Journal of Vibration and Acoustics*, **130**(2), 021001-1 – 021001-8.

Capiez-Lernout, E., and Soize, C. 2008b. Robust design optimization in computational mechanics. *Journal of Applied Mechanics – Transactions of the ASME*, **75**(2), 021001-1 – 021001-11.

Capiez-Lernout, E., and Soize, C. 2008c. Robust updating of uncertain damping models in structural dynamics for low- and medium-frequency ranges. *Mechanical Systems and Signal Processing*, **22**(8), 1774–1792.

Capiez-Lernout, E., Soize, C., Lombard, J.-P., Dupont, C., and Seinturier, E. 2005. Blade manufacturing tolerances definition for a mistuned industrial bladed disk. *Journal of Engineering for Gas Turbines and Power*, **127**(3), 621–628.

Capiez-Lernout, E., Pellissetti, M., Pradlwarter, H. J., Schuëller, G. I., and Soize, C. 2006. Data and model uncertainties in complex aerospace engineering systems. *Journal of Sound and Vibration*, **295**(3–5), 923–938.

Capiez-Lernout, E., Soize, C., and Mignolet, M. P. 2012. Computational stochastic statics of an uncertain curved structure with geometrical nonlinearity in three-dimensional elasticity. *Computational Mechanics*, **49**(1), 87–97.

Chebli, H., and Soize, C. 2004. Experimental validation of a nonparametric probabilistic model of non homogeneous uncertainties for dynamical systems. *Journal of the Acoustical Society of America*, **115**(2), 697–705.

Chen, H., and Chan, C. T. 2007. Acoustic cloaking in three dimensions using acoustic metamaterials. *Applied Physics Letters*, **91**, 183518 (3pp).

Chen, G., and Zhou, J. 1992. *Boundary Element Methods*. New York: Academic Press.

Chen, C., Duhamel, D., and Soize, C. 2006. Probabilistic approach for model and data uncertainties and its experimental identification in structural dynamics: case of composite sandwich panels. *Journal of Sound and Vibration*, **294**(1–2), 64–81.

Cheng, Y., Yang, F., Xu, J. Y., and J., Liu X. 2008. A multilayer structured acoustic cloak with homogeneous isotropic materials. *Applied Physics Letters*, **92**, 151913 (3pp).

Coleman, B. D. 1964. On the thermodynamics, strain impulses and viscoelasticity. *Archive for Rational Mechanics and Analysis*, **17**, 230–254.

Colton, D. L., and Kress, R. 1992. *Integral Equation Methods in Scattering Theory*. Malabar, Florida: Krieger Publishing Company.

Costabel, M., and Stephan, E. 1985. A direct boundary integral equation method for transmission problems. *Journal of Mathematical Analysis and Applications*, **106**, 367–413.

Cotoni, V., Shorter, P. J., and Langley, R. S. 2007. Numerical and experimental validation of a hybrid finite element-statistical energy analysis method. *Journal of the Acoustical Society of America*, **122**(1), 259–270.

Cottereau, R., Clouteau, D., and Soize, C. 2007. Construction of a probabilistic model for impedance matrices. *Computer Methods in Applied Mechanics and Engineering*, **196**(17–20), 2252–2268.

Cremer, L., Heckl, M., and Ungar, E. E. 1988. *Structure-Born Sound*. Berlin: Springer-Verlag.

Crighton, D. G., Dowling, A. P., Ffowcs-Williams, J. E., Heckl, M., and Leppington, F. G. 1992. *Modern Methods in Analytical Acoustics*. Berlin: Springer-Verlag.

Das, S., and Ghanem, R. 2009. A bounded random matrix approach for stochastic upscaling. *Multiscale Modeleling and Simulation. A SIAM Interdisciplinary Journal*, **8**(1), 296–325.

Dautray, R., and Lions, J.-L. 1992. *Mathematical Analysis and Numerical Methods for Science and Technology*. Berlin: Springer-Verlag.

David, J. M., and Menelle, M. 2007. Validation of a modal method by use of an appropriate static potential for a plate coupled to a water-filled cavity. *Journal of Sound and Vibration*, **301**(3–5), 739–759.

Deodatis, G., and Spanos, P. D. 2008. 5th International Conference on Computational Stochastic Mechanics. *Special issue of the Probabilistic Engineering Mechanics*, **23**(2–3), 103–346.

Desceliers, C., Soize, C., and Cambier, S. 2004. Non-parametric – parametric model for random uncertainties in nonlinear structural dynamics: application to earthquake engineering. *Earthquake Engineering and Structural Dynamics*, **33**(3), 315–327.

Desceliers, C., Soize, C., Grimal, Q., Talmant, M., and Naili, S. 2009. Determination of the random anisotropic elasticity layer using transient wave propagation in a fluid-solid multilayer: model and experiments. *Journal of the Acoustical Society of America*, **125**(4), 2027–2034.

Deü, J.-F., and Matignon, D. 2010. Simulation of fractionally damped mechanical systems by means of a Newmark-diffusive scheme. *Computers and Mathematics with Applications*, **59**, 1745–1753.

Deü, J.-F., Larbi, W., and Ohayon, R. 2008. Vibration and transient response for structural acoustics interior coupled systems with dissipative interface. *Computer Methods in Applied Mechanics and Engineering*, **197**(51–52), 4894–4905.

Dovstam, K. 1995. Augmented Hooke's law in frequency domain: three dimensional material damping formulation. *International Journal of Solids and Structures*, **32**(19), 2835–2852.

Duchereau, J., and Soize, C. 2006. Transient dynamics in structures with non-homogeneous uncertainties induced by complex joints. *Mechanical Systems and Signal Processing*, **20**(4), 854–867.

Durand, J.-F., Soize, C., and Gagliardini, L. 2008. Structural-acoustic modeling of automotive vehicles in presence of uncertainties and experimental identification and validation. *Journal of the Acoustical Society of America*, **124**(3), 1513–1525.

Fahy, F. J., and Gardonio, P. 2007. *Sound and Structural Vibration, Second Edition: Radiation, Transmission and Response*. Oxford: Academic Press.

Fang, N., Xi, D., Xu, J., Ambati, M., and Srituravanich, W. 2006. Ultrasonic metamaterials with negative modulus. *Nature materials*, **5**(6), 452–456.

Farhat, C., Harari, I., and Hetmaniuk, U. 2003. A discontinuous Galerkin method with Lagrange multipliers for the solution of Helmholtz problems in the mid-frequency regime. *Computer Methods in Applied Mechanics and Engineering*, **192**(11–12), 1389–1419.

Farhat, C., Wiedemann-Goiran, P., and Tezaur, R. 2004. A discontinuous Galerkin method with plane waves and Lagrange multipliers for the solution of short wave exterior Helmholtz problems on unstructured meshes. *Wave Motion*, **39**(4), 307–317.

Feldman, M. 2011. *Hilbert Transform Applications in Mechanical Vibration*. New York: John Wiley & Sons.

Fernandez, C., Soize, C., and Gagliardini, L. 2009. Fuzzy structure theory modeling of sound-insulation layers in complex vibroacoustic uncertain systems: theory and experimental validation. *Journal of the Acoustical Society of America*, **125**(1), 138–153.

Fernandez, C., Soize, C., and Gagliardini, L. 2010. Sound-insulation layer modelling in car computational vibroacoustics in the medium-frequency range. *Acta Acustica united with Acustica (AAUWA)*, **96**(3), 437–444.

Fishman, G.S. 1996. *Monte Carlo: Concepts, algorithms, and applications*. New York: Springer-Verlag.

Fung, Y. C. 1968. *Foundations of Solid Mechanics*. Englewood Cliffs, New Jersey: Prentice Hall.

Gaul, L., Kögl, M., and Wagner, M. 2003. *Boundary Element Methods for Engineers and Scientists*. Heidelberg, New York: Springer-Verlag.

Geers, T. L., and Felippa, C. A. 1983. Doubly asymptotic approximations for vibration analysis of submerged structures. *Journal of the Acoustical Society of America*, **173**(4), 1152–1159.

Geradin, M., and Rixen, D. 1997. *Mechanical Vibrations, Second Edition: Theory and Applications to Structural Dynamics*. Chichester: Wiley.

Ghanem, R., and Sarkar, A. 2003. Reduced models for the medium-frequency dynamics of stochastic systems. *Journal of the Acoustical Society of America*, **113**(2), 834–846.

Ghanem, R., and Spanos, P. D. 1991. *Stochastic Finite Elements: a Spectral Approach*. New York: Springer-Verlag.

Ghanem, R., and Spanos, P. D. 2003. *Stochastic Finite Elements: A Spectral Approach*. New York: Dover Publications.

Givoli, D. 1992. *Numerical Methods for Problems in Infinite Domains*. Amsterdam, London, New York, Tokyo: Elsevier.

Golla, D. F., and Hughes, P. C. 1985. Dynamics of viscoelastic structures: a time domain, finite element formulation. *Journal of Applied Mechanics – Transactions of the ASME*, **52**, 897–906.

Goller, B., Pradlwarter, H. J., and Schuëller, G. I. 2009. Robust model updating with insufficient data. *Computer Methods in Applied Mechanics and Engineering*, **198**(37–40), 3096–3104.

Golub, G. H., and Van Loan, C. F. 1989. *Matrix Computations*. Baltimore and London: The Johns Hopkins University Press.

Greengard, L., and Rokhlin, V. 1987. A fast algoritm for particle simulations. *Journal of Computational Physics*, **73**(2), 325–348.

Gumerov, N. A., and Duraiswami, R. 2004. *Fast Multipole Methods for the Helmholtz Equation in Three Dimension*. Amsterdam: Elsevier.

Gurtin, M. E., and Herrera, I. 1965. On dissipation inequalities and linear viscoelasticity. *Quarterly of Applied Mathematics*, **23**(3), 235–245.

Hackbusch, W. 1995. *Integral Equations, Theory and Numerical Treatment*. Basel: Birkhauser Verlag.

Hahn, S. L. 1996. *Hilbert transforms in signal processing*. Boston: Artech House Signal Processing Library.

Harari, I., Grosh, K., Hughes, T. J. R., Malhotra, M., Pinsky, P. M., Stewart, J. R., and Thompson, L. L. 1996. Recent development in finite element methods for structural acoustics. *Archives of Computational Methods in Engineering*, **3**(2–3), 131–309.

Howe, M. S. 2008. *Acoustics of Fluid-Structure Interactions*. Cambridge Monographs on Mechanics. Cambridge: Cambridge University Press.

Hsiao, G. C., and Wendland, W. L. 2008. *Boundary Integral Equations*. Berlin, Heidelberg: Springer-Verlag.

Hughes, T. J. R. 2000. *The Finite Element Method: Linear Static and Dynamic Finite Element Analysis*. New York: Dover Publications.

Hughes, T. J. R., Cottrell, J. A., and Bazilevs, Y. 1996. Isogeometric analysis: CAD, finite elements, NURBS, exact geometry and mesh refinement. *Computer Methods in Applied Mechanics and Engineering*, **194**, 4135–4195.

Jaynes, E. T. 1957. Information theory and statistical mechanics. *Physical Review*, **108**(2), 171–190.

Jentsch, L., and Natroshvili, D. 1999. Non-local approach in mathematical problems of fluid-structure interaction. *Mathematical Methods in Applied Sciences*, **22**(1), 13–42.

Jones, D. S. 1974. Integral equations for the exterior acoustic problem. *Quarterly Journal of Mechanics and Applied Mathematics*, **1**(27), 129–142.

Jones, D. S. 1983. Low-frequency scattering by a body in lubricated contact. *Quarterly Journal of Mechanics and Applied Mathematics*, **36**, 111–137.

Jones, D. S. 1986. *Acoustic and Electromagnetic Waves*. New York: Oxford University Press.

Junger, M. C., and Feit, D. 1993. *Sound, Structures and Their Interaction*. Woodbury, NY: Acoust. Soc. Am. Publications on Acoustics. (Originally published in 1972, MIT Press, Cambridge.)

Kaipio, J., and Somersalo, E. 2005. *Statistical and Computational Inverse Problems*. New York: Springer-Verlag.

Kassem, M., Soize, C., and Gagliardini, L. 2009. Energy density field approach for low- and medium-frequency vibroacoustic analysis of complex structures using a stochastic computational model. *Journal of Sound and Vibration*, **323**(3–5), 849–863.

Kassem, M., Soize, C., and Gagliardini, L. 2011. Structural partitioning of complex structures in the medium-frequency range: an application to an automotive vehicle. *Journal of Sound and Vibration*, **330**(5), 937–946.

King, F. W. 2009. *Hilbert Transforms*. Vol 1 and Vol 2. Encyclopedia of Mathematics and its Applications. Cambridge: Cambridge University Press.

Kramers, H. A. 1927. La diffusion de la lumière par les atomes. Pages 545–557 of: *Atti del Congresso Internazionale dei Fisica*, vol. 2. Como, Italy: Transactions of Volta Centenary Congress.

Kress, R. 1989. *Linear Integral Equations*. New York: Springer.

Kronig, R. D. 1926. On the theory of dispersion of X-rays. *Journal of Optical Society of America*, **12**(6), 547–557.

Landau, L., and Lifchitz, E. 1992. *Fluid Mechanics*. Oxford: Pergamon Press.

Le Maître, O.P., and Knio, O.M. 2010. *Spectral Methods for Uncerainty Quantification with Applications to Computational Fluid Dynamics*. Heidelberg: Springer.

Lee, M., Park, Y. S., Park, Y., and Park, K. C. 2009. New approximations of external acoustic-structural interactions: derivation and evaluation. *Computer Methods in Applied Mechanics and Engineering*, **198**(15–16), 1368–1388.

Lesieutre, G. A. 2010. Damping in structural dynamics. In: Blockley, R., and Shyy, W. (eds), *Encyclopedia of Aerospace Engineering*. New York: John Wiley & Sons. doi: 10.1002/9780470686652.eae146

Lesieutre, G. A., and Mingori, D.L. 1990. Finite element modeling of frequency-dependent material damping using augmenting thermodynamic fields. *Journal of Guidance, Control, and Dynamics*, **13**, 1040–1050.

Lighthill, J. 1978. *Waves in Fluids*. Cambridge: Cambridge University Press.

Luke, C. J., and Martin, P. A. 1995. Fluid-solid interaction: acoustic scattering by a smooth elastic obstacle. *SIAM Journal on Applied Mathematics*, **55**(4), 904–922.

Mace, B., Worden, W., and Manson, G. 2005. Uncertainty in structural dynamics. *Special issue of the Journal of Sound and Vibration*, **288**(3), 431–790.

Mathews, I. C. 1986. Numerical techniques for three-dimensional steady-state fluid-structure interaction. *Journal of the Acoustical Society of America*, **79**, 1317–1325.

Mc Tavish, D. J., and Hughes, P. C. 1993. Modeling of linear viscoelastic space structures. *Journal of Vibration and Acoustics*, **115**, 103–113.

Mehta, M. L. 1991. *Random Matrices, Revised and Enlarged Second Edition*. New York: Academic Press.

Mignolet, M. P., and Soize, C. 2008a. Nonparametric stochastic modeling of linear systems with prescribed variance of several natural frequencies. *Probabilistic Engineering Mechanics*, **23**(2–3), 267–278.

Mignolet, M. P., and Soize, C. 2008b. Stochastic reduced order models for uncertain nonlinear dynamical systems. *Computer Methods in Applied Mechanics and Engineering*, **197**(45–48), 3951–3963.

Mignolet, M. P., Soize, C., and Avalos, J. 2013. Nonparametric stochastic modeling of structures with uncertain boundary conditions\coupling between substructures. *AIAA Journal*, **51**(6), 1296–1308.

Milton, G. W., Briane, M., and Willis, J. R. 2006. On cloaking for elasticity and physical equations with a transformation invariant form. *New Journal of Physics*, **8**, 248 (20pp).

Morand, H. J.-P., and Ohayon, R. 1995. *Fluid Structure Interaction*. Chichester: John Wiley & Sons.

Morse, P. M., and Ingard, K. U. 1968. *Theoretical Acoustics*. New York: McGraw-Hill.

Nedelec, J. C. 2001. *Acoustic and Electromagnetic Equations. Integral representation for harmonic Problems*. New York: Springer.

Oden, J. T., Prudhomme, S., and Demkowicz, L. 2005. A posteriori error estimation for acoustic wave propagation. *Archives of Computational Methods in Engineering*, **12**(4), 343–389.

Ohayon, R. 2004a. Fluid-structure interaction problems. Pages 683–693 of: Stein, E., de Borst, R., and Hughes, T.J.R. (eds), *Encyclopedia of Computational Mechanics*. Vol. 2: Solids and Structures. Chichester: John Wiley & Sons.

Ohayon, R. 2004b. Reduced models for fluid-structure interaction problems. *International Journal of Numerical Methods in Engineering*, **60**(1), 139–152.

Ohayon, R., and Sanchez-Palencia, E. 1983. On the vibration problem for an elastic body surrounded by a slightly compressible fluid. *R.A.I.R.O. Analyse numrique/Numerical Analysis*, **17**(3), 311–326.

Ohayon, R., and Soize, C. 1998. *Structural Acoustics and Vibration*. London: Academic Press.

Ohayon, R., and Soize, C. 2012. Advanced computational dissipative structural acoustics and fluid-structure interaction in low- and medium-frequency domains. *International Journal of Aeronautical and Space Sciences*, **13**(2), 14–40.

Ohayon, R., Sampaio, R., and Soize, C. 1997. Dynamic substructuring of damped structures using singular value decomposition. *Journal of Applied Mechanics – Transactions of the ASME*, **64**(2), 292–298.

Pandey, J. N. 1996. *The Hilbert Transform of Schwartz Distributions and Applications*. New York: John Wiley & Sons.

Panich, O. I. 1965. On the question of solvability of the exterior boundary value problems for the wave equation and Maxwell's equations. *Russian Math. Surv.*, **20**, 221–226.

Papadimitriou, C., Beck, J. L., and Katafygiotis, L. S. 2001. Updating robust reliability using structural test data. *Probabilistic Engineering Mechanics*, **16**(2), 103–113.

Papoulis, A. 1977. *Signal Analysis*. New York: McGraw-Hill.

Pellissetti, M., Capiez-Lernout, E., Pradlwarter, H. J., Soize, C., and Schuëller, G. I. 2008. Reliability analysis of a satellite structure with a parametric and a non-parametric probabilistic model. *Computer Methods in Applied Mechanics and Engineering*, **198**(2), 344–357.

Pendry, J. B., and Li, J. 2008. An acoustic metafluid: realizing a broadband acoustic cloak. *New Journal of Physics*, **10**(2), 115032 (9pp).

Pierce, A. D. 1989. *Acoustics: An Introduction to its Physical Principles and Applications*. Woodbury, NY: Acoust. Soc. Am. Publications on Acoustics. (Originally published in 1981, McGraw-Hill, New York.)

Pinkus, A., and Zafrany, S. 1997. *Fourier Series and Integral Transforms*. Cambridge: Cambridge University Press.

Ritto, T. G., Soize, C., and Sampaio, R. 2009. Nonlinear dynamics of a drill-string with uncertainty model of the bit-rock interaction. *International Journal of Non-Linear Mechanics*, **44**(8), 865–876.

Ritto, T. G., Soize, C., and Sampaio, R. 2010. Robust optimization of the rate of penetration of a drill-string using a stochastic nonlinear dynamical model. *Computational Mechanics*, **45**(5), 415–427.

Roach, G. F. 1982. *Green's Functions*. 2nd ed. Cambridge: Cambridge University Press.

Rubinstein, R. Y., and Kroese, D. P. 2008. *Simulation and the Monte Carlo Method*. 2nd ed. New York: John Wiley & Sons.

Sampaio, R., and Soize, C. 2007a. On measures of non-linearity effects for uncertain dynamical systems: application to a vibro-impact system. *Journal of Sound and Vibration*, **303**(3–5), 659–674.

Sampaio, R., and Soize, C. 2007b. Remarks on the efficiency of POD for model reduction in nonlinear dynamics of continuous elastic systems. *International Journal for Numerical Methods in Engineering*, **72**(1), 22–45.

Sanchez-Hubert, J., and Sanchez-Palencia, E. 1989. *Vibration and Coupling of Continuous Systems. Asymptotic Methods*. Berlin: Springer-Verlag.

Schanz, M., and Steinbach, O. (Eds.). 2007. *Boundary Element Analysis*. Berlin, Heidelberg, New York: Springer.

Schenck, H. A. 1968. Improved integral formulation for acoustic radiation problems. *Journal of the Acoustical Society of America*, **44**, 41–58.

Schuëller, G. I. 2005. Uncertainties in structural mechanics and analysis-computational methods. *Special issue of Computer and Structures*, **83**(14), 1031–1150.

Schuëller, G. I. 2007. On the treatment of uncertainties in structural mechanics and analysis. *Computer and Structures*, **85**(5–6), 235–243.

Schuëller, G. I., and Jensen, H. A. 2008. Computational methods in optimization considering uncertainties: an overview. *Computer Methods in Applied Mechanics and Engineering*, **198**(1), 2–13.

Shannon, C. E. 1948. A mathematical theory of communication. *Bell System Technology Journal*, **27**(14), 379–423 & 623–659.

Shorter, P. J., and Langley, R. S. 2005. Vibro-acoustic analysis of complex systems. *Journal of Sound and Vibration*, **288**(3), 669–699.

Soize, C. 1986. Probabilistic structural modeling in linear dynamic analysis of complex mechanical systems. I: Theoretical elements. *La Recherche Aérospatiale*, **5**, 23–48 (English edition).

Soize, C. 1993. A model and numerical method in the medium frequency range for vibroacoustic predictions using the theory of structural fuzzy. *Journal of the Acoustical Society of America*, **94**(2), 849–865.

Soize, C. 1998. Estimation of the fuzzy substructure model parameters using the mean power flow equation of the fuzzy structure. *Journal of Vibration and Acoustics*, **120**(1), 279–286.

Soize, C. 2000. A nonparametric model of random uncertainties on reduced matrix model in structural dynamics. *Probabilistic Engineering Mechanics*, **15**(3), 277–294.

Soize, C. 2001. Maximum entropy approach for modeling random uncertainties in transient elastodynamics. *Journal of the Acoustical Society of America*, **109**(5), 1979–1996.

Soize, C. 2003a. Random matrix theory and non-parametric model of random uncertainties. *Journal of Sound and Vibration*, **263**(4), 893–916.

Soize, C. 2003b. Uncertain dynamical systems in the medium-frequency range. *Journal of Engineering Mechanics*, **129**(9), 1017–1027.

Soize, C. 2005a. A comprehensive overview of a non-parametric probabilistic approach of model uncertainties for predictive models in structural dynamics. *Journal of Sound and Vibration*, **288**(3), 623–652.

Soize, C. 2005b. Random matrix theory for modeling uncertainties in computational mechanics. *Computer Methods in Applied Mechanics and Engineering*, **194**(12–16), 1333–1366.

Soize, C. 2008. Construction of probability distributions in high dimension using the maximum entropy principle: applications to stochastic processes, random fields and random matrices. *International Journal for Numerical Methods in Engineering*, **76**(10), 1583–1611.

Soize, C. 2010a. Generalized probabilistic approach of uncertainties in computational dynamics using random matrices and polynomial chaos decompositions. *International Journal for Numerical Methods in Engineering*, **81**(8), 939–970.

Soize, C. 2010b. Random matrices in structural acoustics. Pages 206–230 of: Weaver, R., and Wright, M. (eds), *New Directions in Linear Acoustics: Random Matrix Theory, Quantum Chaos and Complexity*. Cambridge: Cambridge University Press.

Soize, C. 2012a. *Stochastic Models of Uncertainties in Computational Mechanics*. Reston: American Society of Civil Engineers (ASCE).

Soize, C. 2012b. Stochastic models of uncertainties in computational structural dynamics and structural acoustics. Pages 61–113 of: Elishakoff, I., and

Soize, C. (eds), *Nondeterministics Mechanics*. CISM Courses and Lectures (Udine), International Centre for Mechanical Sciences, vol. 539. Wien, NY: Springer.

Soize, C., and Chebli, H. 2003. Random uncertainties model in dynamic substructuring using a nonparametric probabilistic model. *Journal of Engineering Mechanics*, **129**(4), 449–457.

Soize, C., and Poloskov, I. E. 2012. Time-domain formulation in computational dynamics for linear viscoelastic media with model uncertainties and stochastic excitation. *Computers and Mathematics with Applications*, **64**(11), 3594–3612.

Soize, C., Capiez-Lernout, E., Durand, J.-F., Fernandez, C., and Gagliardini, L. 2008a. Probabilistic model identification of uncertainties in computational models for dynamical systems and experimental validation. *Computer Methods in Applied Mechanics and Engineering*, **198**(1), 150–163.

Soize, C., Capiez-Lernout, E., and Ohayon, R. 2008b. Robust updating of uncertain computational models using experimental modal analysis. *AIAA Journal*, **46**(11), 2955–2965.

Spall, J. C. 2003. *Introduction to Stochastic Search and Optimization*. Chichester: John Wiley & Sons.

Taflanidis, A. A., and Beck, J. L. 2008. An efficient framework for optimal robust stochastic system design using stochastic simulation. *Computer Methods in Applied Mechanics and Engineering*, **198**(1), 88–101.

Truesdell, C. 1973. *Encyclopedia of Physics, Vol. VIa/3, Mechanics of Solids III*. Berlin, Heidelberg, New York: Springer-Verlag.

Von Estorff, O., Coyette, J. P., and Migeot, J. L. 2000. Governing formulations of the BEM in acoustics. Pages 1–44 of: Von Estorff, O. (ed), *Boundary Elements in Acoustics: Advances and Applications*. Southampton: WIT Press.

Wang, X. Q., Mignolet, M. P., Soize, C., and Khannav, V. 2011. Stochastic reduced order models for uncertain infinite-dimensional geometrically nonlinear dynamical system. Stochastic excitation cases. Pages 293–302 of: Zhu, W. Q., K., Lin Y., and Cai, G. Q. (eds), *IUTAM Symposium on Nonlinear Stochastic Dynamics and Control*. IUTAM Bookseries. Hangzhou, China: Zhejiang Univ.

Wright, M., and Weaver, R. 2010. *New Directions in Linear Acoustics: Random Matrix Theory, Quantum Chaos and Complexity*. Cambridge: Cambridge University Press.

Zienkiewicz, O. C., and Taylor, R. L. 2005. *The Finite Element Method For Solid And Structural Mechanics*. 6th ed., Amsterdam: Elsevier, Butterworth-Heinemann.

GLOSSARY

E	mathematical expectation
G	Green function associated with the Helmholtz function
$G_{ijkh}(0)$	initial elasticity tensor for viscoelastic material
$G_{ijkh}(t)$	relaxation functions of viscoelastic material
Q	internal acoustic source density
Q_E	external acoustic source density
Z	wall acoustic impedance
$[A]$	reduced dynamic matrix of the internal acoustic fluid
$[A^S]$	reduced dynamic-stiffness matrix of the structure
$[A^Z]$	reduced dynamic matrix associated with the wall acoustic impedance
$[A_{\text{BEM}}]$	reduced matrix of the impedance boundary operator for the external acoustic fluid
$[A_{\text{FSI}}]$	reduced dynamic matrix for the fluid-structure coupled system
$[C]$	reduced coupling matrix between the internal acoustic fluid and the structure
$[D]$	reduced matrix of the internal acoustic fluid
$[D^S]$	reduced damping matrix of the structure
$[I_m]$	$(m \times m)$ identity matrix
$[K]$	reduced matrix of the internal acoustic fluid
$[K^S]$	reduced stiffness matrix of the structure
$[M]$	reduced matrix of the internal acoustic fluid

$[M^S]$	reduced mass matrix of the structure
$[Mat]^T$	transpose of a real or a complex matrix $[Mat]$
$[\mathbb{A}]$	dynamic matrix of the internal acoustic fluid
$[\mathbb{A}^S]$	dynamic-stiffness matrix of the structure
$[\mathbb{A}^Z]$	dynamic matrix associated with the wall acoustic impedance
$[\mathbb{A}_{BEM}]$	matrix of the impedance boundary operator for the external acoustic fluid
$[\mathbb{A}_{FSI}]$	dynamic matrix for the fluid-structure coupled system
$[\mathbb{C}]$	coupling matrix between the internal acoustic fluid and the structure
$[\mathbb{D}]$	matrix of the internal acoustic fluid
$[\mathbb{D}^S]$	damping matrix of the structure
$[\mathbb{K}]$	matrix of the internal acoustic fluid
$[\mathbb{K}^S]$	stiffness matrix of the structure
$[\mathbb{M}]$	matrix of the internal acoustic fluid
$[\mathbb{M}^A]$	added mass matrix of the internal acoustic fluid
$[\mathbb{M}^S]$	mass matrix of the structure
$[\mathbf{A}]$	random reduced dynamic matrix of the internal acoustic fluid
$[\mathbf{A}^S]$	random reduced dynamic-stiffness matrix of the structure
$[\mathbf{A}_{FSI}]$	random reduced dynamic matrix for the fluid-structure coupled system
$[\mathbf{C}]$	random reduced coupling matrix between the internal acoustic fluid and the structure
$[\mathbf{D}]$	random reduced matrix of the internal acoustic fluid
$[\mathbf{D}^S]$	random reduced damping matrix of the structure
$[\mathbf{G}]$	random matrix
$[\mathbf{G}_0]$	random matrix
$[\mathbf{K}]$	random reduced matrix of the internal acoustic fluid
$[\mathbf{K}^S]$	random reduced stiffness matrix of the structure
$[\mathbf{M}]$	random reduced matrix of the internal acoustic fluid
$[\mathbf{M}^S]$	random reduced mass matrix of the structure
$[\mathcal{P}]$	matrix of internal acoustic modes with fixed wall

$[\mathcal{U}]$	matrix of elastic structural modes in vaccuo
$[\mathbf{G}_C]$	random matrix for vibroacoustic coupling
$[\mathbf{G}_{D^s}]$	random matrix for structural damping
$[\mathbf{G}_D]$	random matrix for internal acoustic fluid
$[\mathbf{G}_{K^s}]$	random matrix for structural stiffness
$[\mathbf{G}_K]$	random matrix for internal acoustic fluid
$[\mathbf{G}_{M^s}]$	random matrix for the structural mass
$[\mathbf{G}_M]$	random matrix for internal acoustic fluid
$\mathbb{F}$	vector of discretized acoustic forces
$\mathbb{F}^S$	vector of discretized structural forces
Γ	coupling interface between the structure and the internal acoustic fluid
Γ_E	coupling interface between the structure and the external acoustic fluid, equal to $\partial\Omega_E$
Γ_Z	coupling interface between the structure and the internal acoustic fluid with acoustical properties
Ω	internal acoustic fluid domain
Ω_E	external acoustic domain
Ω_S	structural domain
Ω_i	$\mathbb{R}^3 \backslash (\Omega_E \cup \Gamma_E)$
Ω_{damp}	viscoelastic part of Ω_S with viscoelastic constitutive equation
Ω_{visco}	damped part of Ω_S with dissipative constitutive equation
$\mathbb{P}$	vector of internal acoustic pressure DOF
$\mathbb{P}_\alpha$	internal acoustic mode with fixed wall
$\mathbb{R}^3$	three-dimensional Euclidean space
$\mathbf{SG}_\varepsilon^+$	ensemble of random matrices
$\mathbb{U}$	vector of structural displacement DOF
$\mathbb{U}_\alpha$	elastic structural mode α in vaccuo
$\|\mathcal{M}\|_F$	Frobenius norm of matrix $[\mathcal{M}]$
$\mathbf{G}$	mechanical surface force field on $\partial\Omega_S$
$\mathbf{I}$	identity operator
$\mathbf{P}$	random vector of internal acoustic pressure DOF

$\mathbf{Q}$	random vector of the generalized coordinates for the internal acoustic fluid
$\mathbf{Q}^S$	random vector of the generalized coordinates for the structure
$\mathbf{U}$	random vector of structural displacement DOF
$\mathbf{Z}_{\Gamma_E}$	impedance boundary operator of external acoustic fluid
$\mathbf{Z}_{\mathrm{rad}}$	radiation impedance operator
$\mathbf{f}$	vector of the generalized forces for the internal acoustic fluid
$\mathbf{f}^S$	vector of the generalized forces for the structure
$\mathbf{g}$	mechanical body force field in the structure
$\mathbf{n}$	outward unit normal to $\partial\Omega$
$\mathbf{n}^S$	outward unit normal to $\partial\Omega_S$
∇	gradient operator
$\nabla_{\mathbf{x}}$	gradient operator with respect to $\mathbf{x}$
$\mathbf{q}$	vector of the generalized coordinates for the internal acoustic fluid
$\mathbf{q}^S$	vector of the generalized coordinates for the structure
$\mathbf{u}$	structural displacement field
$\mathbf{v}$	internal acoustic velocity field
$\mathbf{x}$	generic point of $\mathbb{R}^3$
χ	frequency dependent damping coefficient for parameterized family of structural damping models
$\mathcal{E}$	entropy from information theory
δ	dispersion parameter
ω	circular frequency in rad/s
$\partial\Omega$	boundary of Ω
$\partial\Omega_E$	boundary of Ω_E, equal to Γ_E
$\partial\Omega_S$	boundary of Ω_S
p_{given}	given external acoustic pressure field
$p_{\mathrm{given}}|_{\Gamma_E}$	value of the given external acoustic pressure field on Γ_E
ρ_0	mass density of the internal acoustic fluid
ρ_E	mass density of the external acoustic fluid

ρ_S	mass density of the structure
σ	stress tensor in the structure
σ_{ij}	component of the stress tensor in the structure
$\sigma_{ij}^{\text{elas}}$	component of the elastic stress tensor in the structure
τ	damping coefficient of the internal acoustic fluid
$\times$	vector product
ε_{kh}	component of the strain tensor in the structure
φ	displacement potential field of the internal acoustic fluid
a_{ijkh}	elastic coefficients of the structure
b_{ijkh}	damping coefficients of the structure
c_0	speed of sound in the internal acoustic fluid
c_E	speed of sound in the external acoustic fluid
i	imaginary complex number i
k	wave number in the external acoustic fluid
n	number of internal acoustic DOF
n_j^S	component of vector $\mathbf{n}^S$
n_S	number of structural DOF
n_j	component of vector $\mathbf{n}$
p	internal acoustic pressure field
p_E	external acoustic pressure field
$p_E|_{\Gamma_E}$	value of the external acoustic pressure field on Γ_E
p_X	probability density function of random variable X
$p_{[\mathbf{G}_0]}$	probability density function of random matrix $[\mathbf{G}_0]$
s_{ij}^{damp}	component of the damping stress tensor in the structure
t	time
x_j	coordinate of point $\mathbf{x}$
BEM	boundary element method
BIE	boundary integral equation
DAA	doubly asymptotic approximation
DOF	degrees-of-freedom

DtN   Dirichlet to Neumann

FEM   finite element method
FMM   fast multipole method

GOE   Gaussian orthogonal ensemble

HF   high frequency

LF   low frequency

MF   medium frequency

NRBC   nonlocal nonreflecting boundary condition

ROM   reduced-order model

UQ   uncertainty quantification

AUTHOR INDEX

SUBJECT INDEX